COLLINS
DIY GUIDE

WEATHERPROOFING &
INSULATION

WEATHERPROOFING &
INSULATION
JACKSON & DAY

HarperCollins*Publishers*

Published by HarperCollins Publishers
London

This book was created exclusively for
HarperCollins Publishers by
Jackson Day Jennings Ltd trading as
Inklink.

**Design, art direction
and project management**
Simon Jennings

Text
Albert Jackson
David Day

**Text and
editorial direction**
Albert Jackson

Illustrations editor
David Day

**Designer and
production assistant**
Alan Marshall

Illustrators
David Day
Robin Harris

Additional illustrations
Brian Craker
Michael Parr
Brian Sayers

First published in 1988
This edition published 1992,
reprinted in 1993, 1994

The text and illustrations in this book
were previously published in
Collins Complete DIY Manual

Copyright © 1986, 1988, 1992
HarperCollins Publishers

ISBN 0 00 412816 8

A catalogue record for this book is
available from the British Library

Printed and bound in Hong Kong

Picture credits
Blue Circle Industries PLC: 35B: 49
Cement & Concrete Association: 39
Magnet & Southerns: 20
Paul Chave: 16
Peter Higgins: 45
Rentokil: 15R
Simon Jennings: 10, 15L, 16, 32, 35TC, 41

CONTENTS

Cross-references
There are few DIY projects that do not require a combination of skills. Decorating a single room, for instance, might also involve modifying the plumbing or electrical wiring, installing ventilation or insulation, repairing the structure of the building and so on. As a result, you might have to refer to more than one section of this book. To help you locate the relevant sections, a symbol ($\triangleright$) in the text refers you to a list of cross-references in the page margin. Those references printed in bold type are directly related to the task in hand. Other references which will broaden your understanding of the subject are printed in light-weight type.

DRAUGHTPROOFING DOORS AND WINDOWS

Seal off the major draughts around windows and doors. For a modest outlay, draughtproofing provides a substantial return both economically and in terms of your comfort.

THRESHOLD DRAUGHT EXCLUDERS

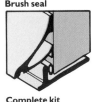

Flexible strip   Brush seal

Flexible arch   Complete kit

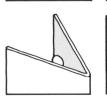
Automatic excluder

DRAUGHTPROOFING DOORS

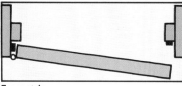
Foam strip

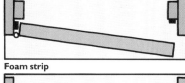
Flexible tube

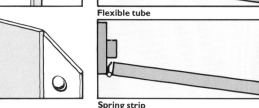
Spring strip

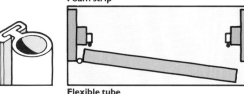
V-strip

SASH WINDOW EXCLUDERS

Spring or V-strip

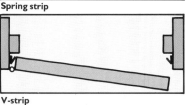

Tubular strip
This is removed in warm weather.

Brush seal
The sashes slide past the brush seal.

Threshold draught excluders

The gap between the door and floor can be very large and will admit fierce draughts. Use a threshold excluder to seal this gap.

Flexible-strip excluder
The simplest form of threshold excluder is a flexible strip of plastic or rubber which sweeps against the floorcovering to form a seal. This type of excluder is rarely suitable for exterior doors and quickly wears out. However, it is inexpensive and easy to fit.

Brush seal
A nylon-bristle brush set into a metal or plastic extrusion acts as a draught excluder. It is suitable for slightly uneven or textured floorcoverings; the same excluder works on both hinged and sliding doors.

Automatic excluder
A plastic strip and its extruded clip are spring-loaded to lift from the floor as the door is opened. On closing the door, the excluder is pressed against the floor by a stop screwed to the frame.

Flexible-arch excluder
An aluminium extrusion with a vinyl arched insert, which presses against the bottom edge of the door. If you fit one for an external door, make sure it is fitted with under-seals to stop rain seeping beneath it.

Complete door kit
The best solution for an outside door is a kit combining an aluminium weather trim, which sheds the rainwater, and a weather bar with a built-in tubular draught excluder screwed to the floor.

Weatherstripping door edges

Any well-fitting door requires a gap of 2mm (1/16in) at top and sides so that it can be operated smoothly. However, the combined area of a gap this large loses a great deal of heat.

Foam strip
The most straightforward excluder is a self-adhesive foam-plastic strip. It is compressed by the door, forming a seal. Don't stretch foam excluders as it reduces their efficiency.

Flexible-tube excluder
A small vinyl tube held in a plastic or metal extrusion is compressed to fill the gap around the door.

Spring-strip excluder
Thin metal or plastic strips with a sprung leaf are pinned or glued to the door frame. The top and closing edges of the door brush past the leaf, sealing the gap. The hinged edge compresses it.

V-strip excluder
A variation on the spring strip, the leaf is bent right back to form a V-shape.

Draughtproofing sealant
With this excluder, a bead of flexible sealant is squeezed onto the door stop: a low-tack tape applied to the surface of the door acts as a release agent. When the door is closed, it flattens the bead, which fills the gap perfectly. When it is set, the parting layer of tape is peeled from the door, leaving the sealant firmly attached to the door frame. If the door warps, the seal is not so good.

Sealing a window

Hinged casement or pivot windows can be sealed with any of the draught excluders suggested for fitting around the edge of a door, but draughtproofing a sliding sash window presents a more difficult problem.

The top and bottom closing rails of a sash window can be sealed with any form of compressible excluder; the sliding edges admit fewer draughts but they can be sealed with a brush seal fixed to the frame, inside for the lower sash, outside for the top one.

A spring or V-strip could be used to seal the gap between the central meeting rails but you may not be able to reverse the sashes once it is fitted. Perhaps the simplest solution is to seal it with a reusable tubular plastic strip.

Clear liquid sealer
If you plan never to open a window during the winter, you could seal all gaps with a clear liquid draught seal, applied from a tube. It is virtually invisible when dry and can be peeled off, without damaging the paintwork, when you want to open the window again.

Damp, or rather the symptoms of it, can be most distressing both in terms of your health and the condition of your home. Try to locate the source of the problem as quickly as possible before it promotes its even more damaging side effects – wet and dry rot. Unfortunately, this is sometimes easier said than done, as one form of damp may obscure another, or may appear in an unfamiliar guise. The two main classifications are penetrating and rising damp, although condensation may appear as one or the other.

Principal causes of penetrating damp
1 Broken gutter
2 Leaking downpipe
3 Missing roof tile
4 Damaged flashing
5 Faulty pointing
6 Porous brick
7 Cracked masonry
8 Cracked render
9 Blocked drip groove
10 Defective seals around frames
11 Missing weatherboard
12 Bridged cavity

Principal causes of rising damp
● Missing DPC or DPM
● Damaged DPC or DPM
● DPC too low
● Bridged DPC
● Earth piled above DPC

SEE ALSO

Details for: ▷
Wet and dry rot   15–16

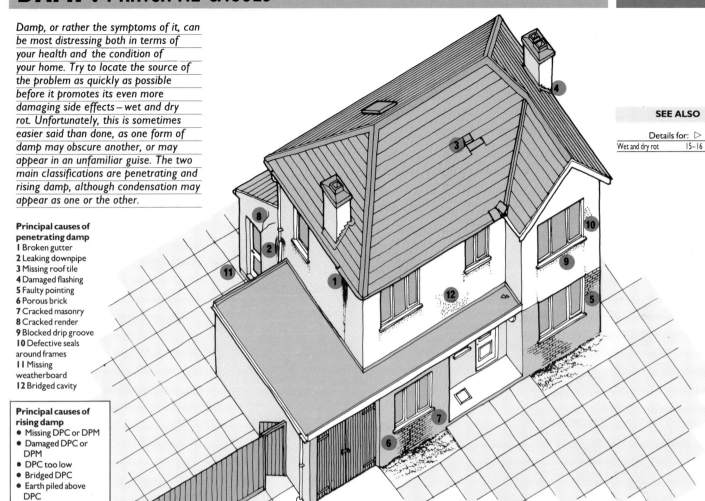

Penetrating damp

Penetrating damp is the result of water permeating the structure of the house from outside. The symptoms occur with wet weather only. After a few dry days, damp patches dry out but frequently leave stains.

Isolated patches are caused by a heavy deposit of water in one area and should pinpoint the source fairly accurately. General dampness usually indicates that the wall itself has become porous, but it could equally be caused by some other problem.

Penetrating damp occurs more often in older homes with solid walls. Relatively modern houses built with a cavity between two thinner brick skins are less likely to suffer from penetrating damp, unless the cavity is bridged in one of several ways.

Rising damp

Rising damp is caused by water soaking up from the ground into the floors and walls of the house. Most houses are protected with an impervious barrier built into the walls and under concrete floors so that water cannot permeate above a certain level.

If the damp-proof course (DPC) in the walls or the membrane (DPM) in a floor breaks down, water leaks into the upper structure. Alternatively, there may be something forming a bridge across the barrier so that water is able to flow around it. Some older houses were built without a DPC.

Rising damp is confined to solid floors and the lower sections of walls. It is a constant problem even in dry weather but becomes worse with prolonged wet weather.

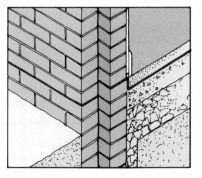

DPC in a solid wall
A layer of impervious material is built into a joint between brick courses, 150mm (6in) above the ground.

DPC and DPM in a cavity wall structure
The damp-proof membrane in a concrete floor is linked to the DPC protecting the inner leaf of the wall. The outer leaf has its own damp-proof course.

PENETRATING DAMP: PRINCIPAL CAUSES

CAUSE	SYMPTOMS	REMEDY
Broken or blocked gutter Rainwater overflows the gutter, typically at the joints of old metal types, and saturates the wall directly below, so that it is prevented from drying out normally.	Damp patches appearing near the ceiling in upstairs rooms. Mould forming immediately behind the leak.	Clear leaves and silt from the gutters. Repair the gutters, or replace a faulty system with a maintenance-free plastic set-up (◁).
Broken or blocked downpipes A downpipe that has cracked or rusted douses the wall immediately behind the leak. Leaves lodged behind the pipe at the fixing brackets will produce a similar effect eventually.	An isolated damp patch, often appearing halfway up the wall. Mould growth behind the pipe.	Repair or replace the defective downpipe, using a maintenance-free plastic type (◁). Clear the blockage.
Loose or broken roof tiles Defective tiles allow rainwater to penetrate the roof.	Damp patches appearing on upstairs ceilings, usually during a heavy downpour.	Replace the faulty tiles (◁), renewing any damaged roofing felt.
Damaged flashing The joins between the roof of a lean-to extension and the side wall of the house, or where a chimney stack emerges from the roof, are sealed with flashing strips, usually lead or zinc (or sometimes a mortar fillet). When the flashing cracks, peels or parts from its fixing, water trickles down the wall or down the chimney stack.	Damp patch on the ceiling extending from the wall or chimney breast; also on the chimney breast itself. Damp patch on the side wall near the join with the lean-to extension; damp patch on the lean-to ceiling itself.	Repair the existing flashing by refitting if it appears to be undamaged, or replace it with a similar material or a self-adhesive flashing strip (◁).
Faulty pointing Ageing mortar between bricks will eventually dry and fall out; water then penetrates the remaining jointing mortar to the inside of the wall.	Isolated damp patches or sometimes widespread dampness, depending on the extent of the deterioration.	Repoint the joints between bricks (◁) then treat the entire wall with water-repellent fluid or paint.
Porous bricks Bricks in good condition are weatherproof but old, soft bricks become porous and often lose their faces so that the whole wall is eventually saturated, particularly on an elevation that faces prevailing winds, or where some other drainage fault occurs.	Widespread damp on the inner face of exterior walls. A noticeable increase in damp during a downpour. Mould growth appearing on internal plaster and decorations.	Waterproof the exterior with a clear repellent fluid or exterior paint, or cement-render the surface (◁) where the deterioration is extensive.
Cracked brickwork A crack in a brick wall allows rainwater (or water from a leak) to seep inside then run to the inside face.	An isolated damp patch, which may appear on a chimney breast if the stack is cracked.	Fill the cracks and replace any damaged brickwork (◁).
Defective render Cracked or blown render encourages rainwater to seep between it and brick wall behind. The water is prevented from evaporating and so becomes absorbed by the wall.	An isolated damp patch, which may become widespread. The trouble can persist for some time after rain ceases.	Fill and reinforce the crack. Hack off the damaged or blown render, patch it with new sand-cement render, then weatherproof the wall by applying exterior paint (◁).
Damaged coping If the coping stone on top of a roof parapet wall (◁) is missing, or the joints are open, water can penetrate the wall.	Damp patches on the ceiling against the wall just below the parapet.	Bed a new stone on fresh mortar and make good the joints.

PENETRATING DAMP: PRINCIPAL CAUSES

CAUSE	SYMPTOMS	REMEDY
Blocked drip groove Exterior window sills should have a groove running longitudinally on the underside. When rainwater runs under, it falls off at the groove before reaching the wall. If the groove becomes bridged with layers of paint or moss, water will soak the wall behind.	Damp patches along the underside of a window frame. Rotting wooden sill on the inside and outside. Mould growth appearing on the inside face of the wall below the window.	Rake out the drip groove. Nail a batten to the underside of a wooden sill to form a deflection for drips (▷).
Failed seals around windows and door frames Timber frames shrink, pulling the pointing from around the edge so that rainwater can penetrate the gap.	Damp surrounding frames and rotting woodwork. Sometimes the gap itself is obvious where the mortar has fallen out.	Repair the frames (▷). Seal around the edge with mastic (▷).
No weatherboard An angled weatherboard across the bottom of a door should shed water clear of the threshold and prevent water running under the door.	Damp floorboards just inside the door. Rotting at the base of the door frame.	Fit a weatherboard even if there are no obvious signs of damage. (▷). Repair the frame (▷).
Bridged wall cavity During building, mortar inadvertently dropped onto a wall tie connecting the inner and outer leaves of a cavity wall allows water to bridge the gap.	An isolated damp patch appearing anywhere on the wall, particularly after a heavy downpour.	Open up the wall and remove the mortar droppings (▷), then waterproof the wall externally with paint or clear repellent (▷).

SEE ALSO

Details for: ▷

Waterproofing bricks	34
New DPM	14, 30
Weatherboard	27
Repairing frames	17–18, 25
Drip batten	10
Sealing frames	10
Bridged cavity	10
New DPC	12–13

RISING DAMP: PRINCIPAL CAUSES

CAUSE	SYMPTOMS	REMEDY
No DPC or DPM If a house was built without either a damp-proof course or damp-proof membrane, water is able to soak up from the ground.	Widespread damp at skirting level. Damp concrete floor surface.	Fit a new DPC or DPM (▷).
Broken DPC or DPM If either the DPC or DPM has deteriorated, water will penetrate at that isolated point.	Possibly isolated but spreading damp at skirting level.	Repair or replace the DPC or DPM (▷).
DPC too low The DPC may not be the necessary 150mm (6in) above ground level. Heavy rain is able to splash above the DPC and soak the wall surface.	Damp at skirting level but only where the ground is too high.	Lower the level of the ground outside. If it's a path or patio, cut a 150mm (6in) wide trench and fill with gravel, which drains rapidly.
Bridged DPC Exterior render taken below the DPC, or fallen mortar at the foot of a cavity wall (within the cavity), allows moisture to cross over to the inside.	Widespread damp at, and just above, skirting level.	Hack off render to expose DPC. Remove several bricks and rake out debris from the cavity (▷).
Debris piled against wall A flower bed, rockery or area of paving built against a wall bridges the DPC. Building material and garden refuse left there will do likewise.	Damp at skirting level in area of bridge only, or spreading from that point.	Remove the earth or debris and allow the wall to dry out naturally.

DPC too low

Render bridges DPC

Earth piled over DPC

CURING DAMP

CONDENSATION

Remedies for different forms of damp are suggested throughout the Principal Causes boxes on the previous pages, and you will find detailed instructions for carrying out many of them in other sections of the book, where they contribute to other factors such as heat loss, poor ventilation, and spoiled decoration. The information below supplements those instructions, by providing advice on measures solely to eradicate damp.

1 Water drips to ground

2 A bridged groove

3 Drip moulding

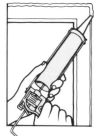

Apply mastic with an applicator gun

WATERPROOFING WALLS

Applying a water repellent to the outside of a wall not only prevents water infusion but also it improves insulation, reducing the possibility of interstitial condensation: this occurs when water vapour from inside the house penetrates the wall until it reaches the damp, colder part of the structure within the brickwork, where it condenses and eventually migrates back to the inner surface, causing stains and mould growth.

There are several damp-proofing liquids for painting on the inside of a wall, but they should be considered a temporary measure only, as they do not cure the source of the problem. Apply two full brush coats over an area appreciably larger than the present extent of the damp. Once dry, you can decorate the wall as required with paint or a wallcovering.

Alternatively, apply a waterproof laminate or paper. It is hung using standard wallpapering techniques using the manufacturer's own primer and adhesive. However, seams must be lapped by 12mm (½in) to prevent moisture penetration. Apply paint or paper over the laminate, or better still, panel it to hide the seams.

PROVIDING A DRIP MOULDING

Because water cannot flow uphill, a drip moulding on the underside of an external window sill forces water to drip to the ground before it reaches the wall behind **(1)**. When decorating, scrape the old paint or moss from the groove before it provides a bridge **(2)**.

You can add a drip moulding to a wooden sill that does not have a pre-cut drip groove by pinning and gluing a 6mm square (¼in square) hardwood strip 35mm (1½in) from the front edge **(3)**. Paint or varnish the strip along with the window sill.

SEALING AROUND WINDOW FRAMES

Scrape out old loose mortar from around the frame. Fill deep gaps with rolled paper or card then seal all round with a flexible mastic. Mastic is available in cartridges to fit an applicator gun, or in tubes, which you squeeze just like toothpaste. Cut the end off the nozzle, then run it down the side of the frame to form a continuous, even bead. If the gap is very wide, fill it with a second bead when the first has set. Most sealants form a skin and can be overpainted after a few hours although they are waterproof even without painting.

BRIDGED CAVITY

The simplest way to deal with a bridged wall cavity which allows water to flow to the inner leaf is to apply a water repellent to the outer surface.

However, this doesn't cure the cause, which may promote other damp conditions later. When it is convenient, when repointing perhaps, remove two or three bricks from the outside in the vicinity of the damp patch by chopping out the mortar around them. Use a small mirror and a torch to inspect the cavity. If you locate mortar lying on a wall tie, chip it off with a rod or opened metal coat hanger and replace the bricks.

Exposing a bridged wall tie
Remove a few bricks to chip mortar from a wall tie.

Air carries moisture as water vapour but its capacity depends on temperature. As it becomes warmer, air absorbs more water like a sponge. When water-laden air comes into contact with a surface that is colder than itself, it cools until it cannot any longer hold the water it has absorbed, and just like the sponge being squeezed, it condenses, depositing water in liquid form onto the surface.

Conditions for condensation

The air in a house is normally warm enough to hold water without reaching saturation point, but a great deal of moisture is also produced by using baths and showers, cooking and even breathing. In cold weather when the low temperature outside cools the external walls and windows below the temperature of the heated air inside, all that extra water runs down window panes and soaks into the wallpaper and plaster. Matters are made worse in the winter by sealing off windows and doors so that fresh air cannot replace humid air before it condenses.

Damp in a fairly new house which is in good condition is almost certainly due to condensation.

The root cause of condensation is rarely simple, as it is a result of a combination of air temperature, humidity, lack of ventilation and thermal insulation. Tackling one of them in isolation may transfer condensation elsewhere or even exaggerate the symptoms. However, the box opposite lists major contributing factors to the total problem.

Condensation appears first on cold glazing

CONDENSATION: PRINCIPAL CAUSES

CAUSE	SYMPTOMS	REMEDY
Insufficient heat The air in an unheated room may already be close to the point of saturation. (Raising the temperature increases the ability of the air to absorb moisture without condensing.)	General condensation.	Heat the room (but not with an oil heater, which produces moisture).
Oil heater An oil heater produces as much water vapour as the paraffin it burns, and condensation will form on windows, walls and ceilings.	General condensation in the room where the heater is used.	Substitute another form of heating.
Uninsulated walls and ceilings Moist air readily condenses on cold exterior walls and ceilings.	Widespread damp and mould. The line of ceiling joists is picked out as mould grows less well along these relatively 'warm' spots.	Install loft insulation (▷) and/or line the ceiling with insulating tiles or polystyrene lining
Cold bridge Even when a wall has cavity insulation, there can be a cold bridge across the lintel over windows and the solid brick down the sides.	Damp patches or mould surrounding the window frames.	Line the walls and window reveal with sheets of expanded polystyrene or foamed polyethylene.
Unlagged pipes Cold water pipes attract condensation. It is often confused with a leak when water collects and drips from the lowest point of a pipe run.	Line of damp on a ceiling or wall following the pipework. Isolated patch on a ceiling, where water drops from plumbing. Beads of moisture on the underside of a pipe.	Insulate the plumbing with proprietary foam lagging tubes or mineral-fibre wrapping (▷).
Cold windows Glass shows condensation usually before any other feature, due to the fact that it's very thin and constantly exposed to the elements.	Misted glass, or water collecting in pools at the bottom of the window pane.	Double-glaze the window (▷). If condensation occurs inside a secondary glazing system, place some silica gel crystals (which absorb moisture) in the cavity between panes (▷).
Sealed fireplace When a fireplace opening is blocked, the air trapped inside the flue cannot circulate and consequently it condenses on the inside, eventually soaking through the brickwork.	Damp patches appearing anywhere on the chimney breast.	Ventilate the chimney by inserting a grille at a low level in the blocked-up part of the fireplace (▷).
Loft insulation blocking airways If loft insulation blocks the spaces around the eaves, air cannot circulate in the roof space, and condensation is able to form.	Widespread mould affecting the timbers in the roof space.	Unblock the airways and, if possible, fit a ventilator grille in the soffit or install tile/slate vents (▷).
Condensation on recent building If you have carried out work involving new bricks, mortar and especially plaster, condensation may be the result of these materials exuding moisture as they dry out.	General condensation affecting walls, ceiling, windows and solid floors.	Wait for the new work to dry out, then review the situation, before decorating or otherwise treating.

SEE ALSO

Details for: ▷	
Lagging pipes	64
Loft insulation	59–60
Wall insulation	62–63
Double glazing	65–68
Secondary glazing	66–68
Chimney ventilation	69
Soffit grille	71
Tile/slate vents	71
Extractor fans	73–76
Dehumidifiers	77

INSERTING A DAMP-PROOF COURSE

When an old damp-proof course (DPC) has failed, or where none exists, the only certain remedy is to insert a new one. Of the four options available, chemical injection is the only method you should attempt yourself. Even then, consider whether it is cost-effective in the long run: rising damp can lead to other expensive repairs unless it is completely eradicated, so hiring a reputable company may prove to be a wise investment (they normally provide a 30-year guarantee). Ask for a detailed specification – known as an Agrément certificate – to ensure that the work is carried out to approved standards and check that the guarantee is covered by insurance in case the company goes out of business.

SEE ALSO

◁ Details for:
Rising damp   7

A physical DPC: a joint is removed to insert impervious layer.

A physical DPC

A traditional DPC consists of a layer of impervious material incorporated into the wall during building at approximately 150mm (6in) – or three brick courses – above ground level. A similar DPC can be installed in an existing building by cutting out a mortar joint with a chain saw or grinding disc. Copper sheet, polythene or bituminous felt is inserted and the joint wedged and filled with fresh mortar. Considerable experience is required to avoid structurally weakening the wall, and there's always the risk of cutting through an electric cable and plumbing pipework. A physical DPC is costly to install but it is considered to be the most reliable method.

Electro-osmosis: a copper electrode is planted in the wall.

Electro-osmosis

This is a method utilizing the principle that a minute electrical charge will prevent water rising by capillary action. Copper electrodes are inserted into the wall and connected to earthing rods buried in the ground. Active systems are connected to the electrical supply but a passive system requires no power at all to operate. These are systems which can only be installed by a professional. Consider for thick walls, which may be difficult to treat by other means.

Porous tubes

Porous clay tubes are inserted into a row of closely spaced holes to increase the rate of evaporation before moisture rises to a higher level. This is a simple and cheap method.

Porous tubes: the wall is drilled to receive the tubes.

The most widely practised method today is to inject a waterproofing chemical, usually silicone-based, to form a continuous barrier throughout the thickness of the wall. It is suitable for brick or stone walls up to 600mm (2ft) thick, and is straightforward to install yourself using hired equipment.

Preparing the wall for injection

If you want to carry out the injection work yourself, use a pressure injection machine, which can be hired. You will require 68 to 90 litres (15 to 20 gallons) of DPC fluid per 30m (100ft) of 225mm (9in) thick wall.

Remove skirtings and hack off plaster and render to a height of 450mm (1ft 6in) above the line of visible damp. Repair and repoint the brickwork with new mortar (▷).

Drilling the injection holes

Drill a row of holes about 150mm (6in) above external ground level, but below a suspended wooden floor or just above one of solid concrete. If the wall has an old DPC, set the new course just above it, but take care not to puncture it when drilling. Use a masonry drill of about 18 to 25mm (¾in to 1in) in diameter but not less than the injecting nozzle of the machine. If possible, drill an identical row of holes from both sides of a wall 225mm (9in) thick or greater, to provide a continuous DPC.

When drilling a 225mm (9in) solid brick wall, the holes should be at 112mm (4½in) centres, about 25mm (1in) below the upper-edge of a brick course. Angle them downwards slightly. Drill 75mm (3in) deep unless treatment is limited to one side only, when you should drill to a depth of 190mm (7½in). Treat each leaf of a cavity wall separately, drilling to a depth of 75mm (3in) in each one.

If the wall is made of impervious stone, drill into the mortar course around each stone block at the proposed DPC level, spacing the holes every 75mm (3in).

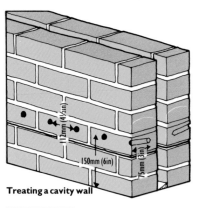

Treating a solid brick wall

Treating a cavity wall

Injecting the fluid

There are various types of injection pump available, but most work in basically the same way. Usually you must insert the pump's filtered suction hose into the drum of chemical. Make sure that the valves controlling the injection nozzles are closed then connect the pump to the mains electrical supply.

Most machines have three nozzles. Connect them to the end of the hoses – if you're treating a thick wall, drill 75mm (3in) deep holes to begin with, and start with the shorter nozzles – and push them into the holes in the wall. Turn their wing nuts to secure them and form a seal (don't overtighten them or you may damage the expansion nipples at their tips). Open the control valves on the two nozzles then switch on the pump so the chemical circulates through the machine.

Bleed off some fluid through the third nozzle into a container to expel air from the system, by opening its valve. Switch it off again and insert in the wall. Re-open the valve and allow the fluid to be injected until it wets the surface of the bricks. Maintain the pressure at about 100 PSI (pounds per square inch) by adjusting the valve on the pump body.

Switch off all three valves then move them on to the next holes and repeat the procedure. When you reach the other end of the wall, switch off the pump then return to the starting point and drill the same holes to 190mm (7½in) deep. Swap the short injection nozzles for the longer 190mm (7½in) ones – you may need to wrap PTFE sealing tape round the threads (▷) – slot them into the wall, tighten their nuts, and inject the fluid.

Flush the machine through with white spirit after use to get rid of the fluid, which otherwise would 'cure' in the system, damaging the pump.

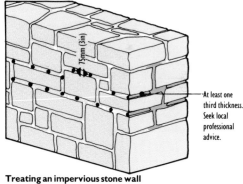

At least one third thickness. Seek local professional advice.

Treating an impervious stone wall

◁ **Hiring the equipment**
Any tool hire specialist will supply you with all the necessary equipment and materials to inject a chemical DPC yourself. It is an economical method requiring careful work rather than experience. Flush the machine thoroughly before you return it.

DAMP-PROOFING A CELLAR

Being at least partially below ground level, the walls and floors of a cellar or basement invariably suffer from damp to some extent. Because the problem cannot be tackled from the outside in the normal way, you will have to seal out the damp by treating the internal surfaces. Rising damp in concrete floors, whatever the situation, can be *treated as described below, but penetrating or rising damp in walls other than in a cellar should be cured at source: merely sealing the internal surface encourages the damp to penetrate elsewhere eventually. In addition, ensure a treated cellar is properly ventilated, and even heated to avoid condensation in the future.*

Treating the floor

If you are laying a new concrete floor, incorporate a damp-proof membrane (DPM) during its construction (◁). If the DPM was omitted or has failed in an existing floor, seal it with a heavy-duty, moisture-curing polyurethane.

Preparing the surface
The floor must be clean and grease-free (◁). Fill any cracks and small holes by priming with one coat of urethane, then one hour later apply a mortar made from 6 parts sand: 1 part cement plus enough urethane to produce a stiff paste. Although urethane can be applied to damp or dry surfaces, it will penetrate a dry floor better, so force-dry excessively damp basements with a fan heater before treatment. Remove all heaters from the room before you begin damp-proofing.

Applying urethane
Use a broom to apply the first coat of urethane using one litre to cover about 5sq m (50sq ft). If you are treating a room with a DPC in the walls, take the urethane coating up behind the skirting to meet it.

Two or three hours later, apply a second coat. Further delay may result in poor inter-coat adhesion. Apply three or four coats in all.

After three days, you can lay any conventional floorcovering or use the floor as it is.

PATCHING ACTIVE LEAKS

Before you damp-proof a cellar, patch cracks which are active water leaks (running water) with a quick-drying hydraulic cement. Supplied as a powder for mixing with water, the cement expands as it hardens, sealing out the running water.

Undercut a crack or hole with a chisel and club hammer. Mix up cement and hold it in the hand until warm then push it into the crack. Hold it in place with your hand or a trowel for three to five minutes until hard.

TREATING THE WALLS

If you want, you can continue with moisture-cured polyurethane to completely seal the walls and floor of a cellar. Decorate with emulsion or oil paints within 24 to 48 hours after treatment for maximum adhesion.

If you'd prefer to hang wallcoverings apply two coats of emulsion paint first and use a heavy-duty paste. Don't hang impervious wallcoverings such as vinyl, however, as it's important that the wall can breathe.

Bitumen latex emulsion
Where you plan to plaster or dry-line the basement walls, you can seal out the damp with a cheaper product, bitumen latex emulsion. It is not suitable as an unprotected covering to walls or floors, although it is often used as an integral DPM under the top screed of a concrete floor and as a waterproof adhesive for some tiles and wooden parquet flooring.

Hack off old plaster to expose the brickwork, then apply a skim coat of mortar to smooth the surface. Paint the wall with two coats of bitumen emulsion, joining with the DPM in the floor. Before the second coat dries, imbed clean, dry sand into it (blinding) to provide a key for the coats of plaster (See below left).

Cement-based waterproof coating
In severe conditions of damp, use a cement-based waterproof coating. Hack off old plaster or render to expose the wall then, to seal the junction between a concrete floor and the wall, cut a chase about 20mm (¾in) wide by the same depth. Brush out the debris and fill the channel with hydraulic cement (see left), finishing it off neatly as an angled fillet.

Mix the powdered cement-based coating to a butter-like consistency, according to the manufacturer's instructions, then apply two coats to the wall with a bristle brush.

However, when brick walls are damp, they bring salts to the surface in the form of white crystals known as efflorescence (◁), so before treating with waterproof coating, apply a salt-inhibiting render consisting of 1 part sulphate-resisting cement: 2 parts clean rendering sand. Add 1 part liquid bonding agent to 3 parts of the mixing water. Apply a thin trowelled coat to a rough wall or brush it onto a relatively smooth surface and allow it to set.

Treating a wall with bitumen latex emulsion
1 Skim coat of mortar
2 Coat of bitumen latex
3 Blinded coat of latex
4 Plaster or dry lining

Moisture-curing polyurethane
Damp-proof a floor with three or four coats of urethane applied with a broom.

DRY AND WET ROT

Rot occurs in unprotected household timbers, fences and outbuildings, which are subjected to damp. Fungal spores, which are always present, multiply and develop in these conditions until eventually the timber is destroyed. Fungal attack can be serious, requiring immediate attention to avoid very costly structural repairs. There are two main scourges: wet and dry rot.

Recognizing rot

Signs of fungal attack are easy enough to detect but it is important to be able to identify certain strains which are much more damaging than others.

Mould growth
White furry deposits or black spots on timber, plaster or wallpaper are mould growths; usually these are a result of condensation. When they are wiped or scraped from the surface, the structure shows no sign of physical deterioration apart from staining. Cure the source of the damp conditions and treat the affected area with a solution of 16 parts warm water: 1 part bleach.

Wet rot

Wet rot occurs in timber with a high moisture content. As soon as the cause is eliminated, further deterioration is arrested. It frequently attacks the framework of doors and windows which have been neglected enabling rainwater to penetrate joints or between brickwork and adjacent timbers. Peeling paintwork is often the first sign, which when removed, reveals timber that is spongy when wet but dark brown and crumbly when dry. In advanced stages, the grain will have split and thin, dark brown fungal strands will be in evidence on the timber. Treat wet rot as soon as practicable.

Dry rot

Once it has taken hold, dry rot is a most serious form of decay. Urgent treatment is essential. It will attack timber with a much lower moisture content than wet rot, but – unlike wet rot, which thrives outdoors as well as indoors – only in poorly ventilated, confined spaces indoors.

Dry rot exhibits various different characteristics depending on the extent of its development. It sends out fine, pale grey tubules in all directions, even through masonry, to seek out and infect other drier timbers: it actually pumps water from damp timber and can progress at an alarming rate. The strands are accompanied by white cotton wool-like growths called mycelium in very damp conditions. When established, dry rot develops wrinkled, pancake-shaped fruiting bodies, which produce rust red spores that are expelled to rapidly cover surrounding timber and masonry. Infested timbers become brown and brittle, exhibiting cracks across and along the grain until it breaks up into cube-like pieces. You may also detect a strong, musty, mushroom-like smell associated with the fungus.

TREATING ROT

Dealing with wet rot
Having eliminated the cause of the damp, cut away and replace badly damaged wood, then paint the new and surrounding woodwork with three liberal applications of fungicidal wood preservative. Brush the liquid well into the joints and end grain.

Before decorating, you can apply a wood hardener to reinforce slightly damaged timbers, then follow six hours later with wood filler to rebuild the surface. Repaint as normal.

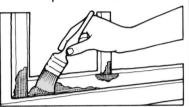

Paint rotted timbers with wood hardener

Dealing with dry rot
Dry rot requires more drastic action and should be treated by a specialist contractor unless the outbreak is minor and self-contained. Remember that dry rot can penetrate masonry: look under the floorboards in adjacent rooms before you are satisfied with the extent of the infection; check cavity walls for signs of rot.

Eliminate the source of water and ensure adequate ventilation in roof spaces or under the floors by unblocking or replacing air bricks. Cut out all infected timber up to at least 450mm (1ft 6in) beyond the last visible sign of rot. Chop plaster from nearby walls, following the strands. Continue for another 450mm (1ft 6in) beyond the extent of the growth. Collect all debris in plastic bags and burn it.

Use a fungicidal preservative fluid to kill remaining spores. Wire brush the masonry then apply three liberal brush-coats to all timber, brickwork and plaster within 1.5m (5ft) of the infected area. Alternatively, hire a coarse sprayer and go over the same area three times.

If a wall was penetrated by strands of dry rot, drill regularly spaced but staggered holes into it from both sides. Angle the holes downwards so that fluid will collect in them to saturate the wall internally. Patch holes after treatment.

Treat replacement timbers and immerse the end grain in a bucket of fluid for five to ten minutes. When you come to make good the wall, you should apply a zinc-oxychloride plaster.

SEE ALSO

Details for: ▷

| Repairing rotten frames | 17–18, 25 |
| Preventing damp | 7–14 |

Wet rot - treat it at your earliest opportunity.

Dry rot - urgent treatment is essential.

Fungal attack can be so damaging that it is well worth taking precautions to prevent it occurring. Regularly decorate and maintain window and door frames, where moisture can penetrate; seal around them with mastic. Provide adequate ventilation between floors and ceilings. Do the same in the loft. Check and eradicate any plumbing leaks and other sources of damp, and you'll be less likely to experience the stranglehold rot can apply.

SEE ALSO

◁ Details for:

| Wet and dry rot | 15 |

Looking after timberwork

Existing and new timbers can be treated with a preservative. Brush and spray two or three applications to standing timbers, paying particular attention to joints and end grain.

Immersing timbers
Timber in contact with the ground would benefit from prolonged immersion in preservative. Stand fence posts on end in a bucket of fluid for 10 minutes. For other timbers, make a shallow bath from loose bricks and line it with thick polythene sheet. Pour preservative into the trough and immerse the timbers, weighing them down with bricks (**1**). To empty the bath, sink a bucket at one end of the trough, then remove the bricks adjacent to it so the fluid pours out (**2**).

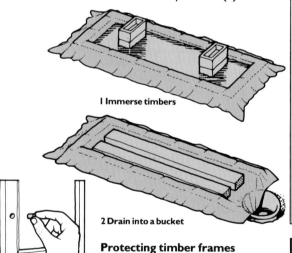

1 Immerse timbers

2 Drain into a bucket

Protecting joints
Place preservative tablets close to the joints of a frame

Protecting timber frames
To protect timber frames, insert preservative in solid tablet form into holes drilled at regular 50mm (2in) intervals in a staggered pattern. If the timber becomes wet, the tablets dissolve, placing preservative exactly where it is needed. Fill the holes with wood filler and paint as normal

WOOD PRESERVATIVES

There are numerous types of wood preservatives for use inside and out.

Choose the correct one for the timber you want to protect:

CLEAR PRESERVATIVE

Clear liquids are specially formulated to protect timber from dry or wet rot. Alternatively, use an all-purpose fluid, which also provides protection against wood-boring insects. Paint or varnish the surface when dry.

COLOURED PRESERVATIVES

Tinted preservatives protect sound timbers against fungal and insect attack and stain the wood at the same time. Some are harmful to plants, however, so check before you buy. There is a choice of brown shades intended to simulate common hardwoods and one specifically for red cedar.

GREEN PRESERVATIVE

Green preservative is normally harmless to plants when it dries, so it is used for horticultural timbers. It can be used to eradicate rot and insect attack as well as protecting sound household timbers. Its colour helps to identify treated timbers in the future. However, the colour is due to its copper content and is not a permanent colouring agent when used outdoors. Its protective properties are unaffected, even when the colour is washed out by heavy rain.

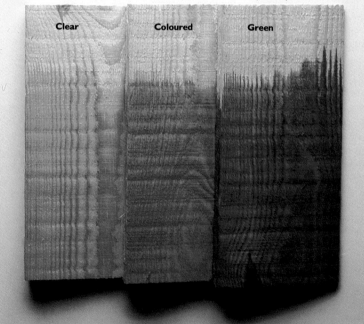

Clear   Coloured   Green

SAFETY WITH PRESERVATIVES

All preservatives are flammable so do not smoke while using, and extinguish naked lights.

    Wear protective gloves at all times when applying preservatives and wear a facemask when using **these liquids indoors.**

    Ensure good ventilation when the liquid is drying, and do not sleep in a freshly treated room for two nights to allow the fumes to dissipate fully.

Softwood is the traditional material for making wooden window frames, and providing it is of sound quality and is well cared for, it will last the life of the building. New frames or frames which have been stripped should always be treated with a clear wood preservative before painting.

Regular maintenance

It is the bottom rail of a softwood window frame that is most vulnerable to rot if it is not protected. The water may be absorbed by the wood through a poor paint finish or by penetrating behind old shrunken putty. An annual check of all window frames should be carried out and any faults should be dealt with. Old putty that has shrunk away from the glass should be cut out and replaced with new (▷).

Remove old flaking paint, make good any cracks in the wood with a flexible filler and repaint, ensuring that the underside of the sash is well painted.

Replacing a sash rail

Where rot is well advanced and the rail is beyond repair it should be cut out and replaced. This should be done before the rot spreads to the stiles of the frame. Otherwise you will eventually have to replace the whole sash frame.

Remove the sash either by unscrewing the hinges or – if it is a double-hung sash window – by removing the beading.

With a little care the repair can be carried out without the glass being removed from the sash frame, though if the window is large it would be safer to take out the glass. In any event, cut away the putty from the damaged rail.

The bottom rail is tenoned into the stiles (1), but it can be replaced by using bridle joints. Saw down the shoulder lines of the tenon joints (2) from both faces of the frame and remove the rail.

Make a new rail, or buy a piece if it is a standard section, and mark and cut it to length with a full-width tenon at each end. Set the positions of the tenons to line up with the mortises of the stiles. Cut the shoulders to match the rebated sections of the stiles (3) or, if it has a decorative moulding, pare the moulding away to leave a flat shoulder (4).

Cut slots in the ends of the stiles to receive the tenons.

Glue the new rail into place with a waterproof resin adhesive and reinforce the two joints with pairs of 6mm (¼in) stopped dowels. Drill the stopped holes from the inside of the frame and stagger them.

When the adhesive is dry, plane the surface as required and treat the new wood with a clear preservative. Re-putty the glass and paint the new rail within three weeks.

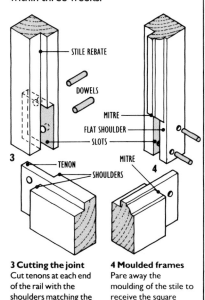

1 The original joint
The rail is tenoned into the stile and fitted with wedges.

2 Cutting out the rail
Saw down the shoulder lines of the joints from both faces of the frame.

3 Cutting the joint
Cut tenons at each end of the rail with the shoulders matching the sections of the stiles.

4 Moulded frames
Pare away the moulding of the stile to receive the square shoulder of the rail. Mitre the moulding.

REPLACING A FIXED-LIGHT RAIL

The frames of some fixed lights are made like sashes but are screwed to the main frame jamb and mullion. Such a frame can be repaired in the same way as a sash (See left) after its glass is removed and it is unscrewed from the window frame. Where this proves too difficult you will have to carry out the repair in situ.

First remove the putty and the glass, then saw through the rail at each end. With a chisel trim the rebated edge of the jamb(s) and/or mullion to a clean surface at the joint (1) and chop out the old tenons. Cut a new length of rail to fit between the prepared edges and cut housings in its top edge at both ends to take loose tenons. Place the housings so that they line up with the mortises and make them twice as long as the depth of the mortises.

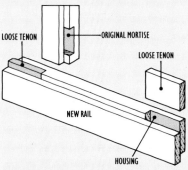

1 Chop out the tenons and cut a new rail to fit

Cut two loose tenons to fit the housings, and two packing pieces. The latter should have one sloping edge (2).

Apply an exterior woodworking adhesive to all of the jointing surfaces, place the rail between the frame members, insert the loose tenons and push them sideways into the mortises. Drive the packing pieces behind the tenons to lock them in place. When the adhesive has set, trim the top edges, treat the new wood with clear preservative, replace the glass and re-putty. Paint within three weeks.

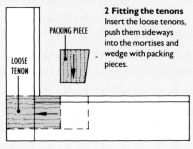

2 Fitting the tenons
Insert the loose tenons, push them sideways into the mortises and wedge with packing pieces.

SEE ALSO

Details for: ▷	
Removing glass	24
Fitting glass	24

● **Removing glass**
Removing glass from a window frame in one piece is not easy so be prepared for it to break. Apply adhesive tape across the glass to bind the pieces together if it should break. Chisel away the putty to leave a clean rebate, then pull out the sprigs (▷). Work the blade of a putty knife into the bedding joint on the inside of the frame to break the grip of the putty. Steady the glass and lift it out when it becomes free.

17

REPAIRING ROTTEN SILLS

The sill is a fundamental part of a window frame, and one attacked by rot can mean major repair work.

A casement window frame is constructed in the same way as a door frame and can be repaired in a similar way (◁). All the glass should be removed first. The window board may also have to be removed, then refitted level with the replacement sill.

Make sure that the damp-proofing of the joint between the underside of the sill and the wall is maintained. Modern gun-applied mastics have made this particular problem easier to overcome. Some traditional frames have a galvanized-iron water bar between the sill and sub-sill. When replacing a sill of this type without removing the whole frame it may be necessary to remove the bar and rely on mastic sealants alone to keep the water out.

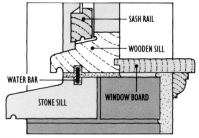

Traditional frame with stone sub-sill

Replacing a wooden sill

If you simply replace a sill by cutting through it and fitting a new section between the jambs you may not have solved your problem. Even when mastic sealants are used, any breakdown of the seal will allow water a direct path to the brickwork and the end grain of the wood, and you may find yourself doing the job all over again.

Serious rot in the sill of a sash window may require the whole frame to be taken out (◁). Make and fit a new sill using the old one as a pattern. Treat the new wood with a preservative and take the opportunity to treat the old wood which is normally hidden by the brickwork. Apply a bead of mastic sealant to the sill, then replace the complete frame in the opening from inside. Make good the plaster.

It is possible to replace the sill from the inside with the frame in place. Saw through the sill close to the jambs and remove the cut centre portion. Cut away the bottom ends of the inner lining level with the pulley stiles and remove the ends of the old sill. Cut the ends of the new sill to fit round the outer lining and under the stiles and inner lining. Fit the sill and nail or screw the stiles to it.

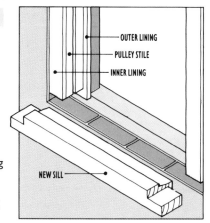

Cut the new sill to fit the frame

Repairing a stone sub-sill

The traditional stone sills that feature in older houses may become eroded by the weather if they are not protected with paint. They may also suffer cracking due to subsidence in part of the wall.

Repair any cracks and eroded surfaces with a quick-setting waterproof cement. Rake the cracks out to clean and enlarge them, then dampen the stone with clean water and work the cement well into the cracks, finishing off flush with the top surface.

Depressions caused by erosion should be undercut to provide the cement with a good hold. A thin layer of cement simply applied to a shallow depression in the surface will not last. Use a cold chisel to cut away the surface of the sill at least 25mm (1in) below the finished level and remove all traces of dust.

Make a wooden former to the shape of the sill and temporarily nail it to the brickwork. Dampen the stone, pour in the cement and tamp it level with the former, then smooth it with a trowel. Leave it to set for a couple of days before removing the former. Let it dry thoroughly before painting (◁).

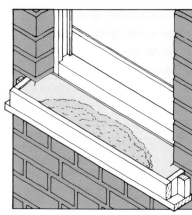

Make a wooden former to the shape of the sill

CASTING A NEW SUB-SILL

Cut out the remains of the old stone sill with a hammer and cold chisel. Make a wooden mould with its end pieces shaped to the same section as the old sill. The mould must be made upside down, its open top representing the underside of the sill.

Fill two thirds of the mould with fine ballast concrete, tamped down well, and then add two lengths of mild steel reinforcing rod, judiciously spaced to share the volume of the sill, then fill the remainder of the mould. Set a narrow piece of wood such as a dowel into notches previously cut in the ends of the mould. This is to form a 'throat' or drip groove in the underside of the sill.

Cover the concrete with polythene sheeting or dampen it regularly for two or three days to prevent rapid drying. When the concrete is set (allow about seven days) remove it from the mould and re-lay the sill in the wall on a bed of mortar to meet the wooden sill.

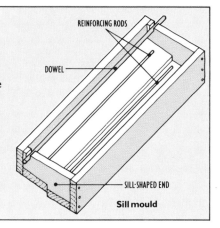

Sill mould

READY-MADE WINDOWS

Joinery suppliers offer a range of ready-made window frames in both hardwood and softwood, and some typical examples are shown below.

Unfortunately the range of sizes is rather limited, but where a ready-made frame is fairly close to one's requirements it is possible to either cut back the brickwork or fill the gap between the frame and the wall with masonry, though this is really acceptable only for windows in rendered walls. In a wall of exposed brickwork the window frame should be made to measure.

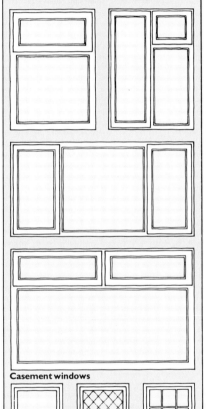

Casement windows

Vertical sliding sash windows

Pivot windows

REPLACEMENT WINDOWS

The style of the windows is an important element in the appearance of any house. Should you be thinking of replacing windows in an older dwelling you might find it better – and not necessarily more expensive – to have new wooden frames made rather than changing to modern windows of aluminium or plastic.

Planning and building regulations

Window conversions do not normally need planning permission as they come under the heading of house improvement or home maintenance, but if you plan to alter your windows significantly – for example, by bricking one up or making a new window opening, or both – you should consult your local Building Control Officer.

All authorities require minimum levels of ventilation to be provided in the habitable rooms of a house, and this normally means that the openable part of windows must have an area at least one twentieth that of the room.

You should also check with your local authority if you live in a listed building or in a conservation area, which could mean some limitation on your choice.

Buying replacement windows

Specialist joinery firms will make up wooden window frames to your size. Specify hardwood or, for a painted finish, softwood impregnated with a timber preservative.

Alternatively you can approach one of the replacement window companies, though this is likely to limit your choice to aluminium or plastic frames. The ready-glazed units can be fitted to your old timber sub-frames or to new hardwood ones supplied by the installer. Most of the replacement window companies operate on the basis of supplying and also fitting the windows, and their service includes disposing of the old windows and of the rubbish.

This method is saving of time and labour, but you should carefully compare the various offerings of these companies and their compatibility with the style of your house before opting for one. Choose a frame that reproduces, as closely as possible, the proportions of the original window.

Replacing a casement window

Measure the width and height of the window opening. If the replacement window will need a timber sub-frame and the existing one is in good condition, take your measurements from inside the frame. Otherwise take them from the brickwork. You may have to cut away some of the rendering or plaster first so as to get accurate measurements. Order the replacement window accordingly.

Remove the old window by first taking out the sashes and then the panes of glass in any fixed part. Unscrew the exposed fixings, such as may be found in a metal frame, or chisel away the plaster or rendering and cut through them with a hacksaw. It should be possible to knock the frame out in one piece, but if not saw through it in several places and lever the pieces out with a crowbar **(1)**. Clean up the exposed brickwork with a bolster chisel to make a neat opening.

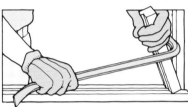

I Lever out the pieces of the old frame

Cut the horns off the new frame if present, then plumb the frame in the window opening and wedge it **(2)**. Drill screw holes through the stiles into the brickwork **(3)**, then remove the frame and plug the holes. Attach a bituminous felt damp-proof course to the stiles and sill and refit the frame, checking again that it is plumb before screwing it into home.

Make good the wall with mortar and plaster. Gaps of 6mm (¼in) or less can be filled with mastic. Glaze the new frame as required.

2 Fit the new frame   3 Drill fixing holes

SEE ALSO

Details for: ▷
Fitting glass   24

19

REPLACEMENT WINDOWS

Bay windows

A bay window is a combination of window frames which are built out from the face of the building. The side frames may be set at 90, 60 or 45 degree angles to the front of the house. Curved bays are also made with equal-sized frames set at a very slight angle to each other to form a faceted curve.

The frames of a bay window are set on brickwork which is built to the shape of the bay, which may be at ground level only, with a flat or pitched roof, or may be continued up through all storeys and finished with a gabled roof.

Bay windows can break away from the main wall through subsidence caused by poor foundations or differential ground movements. Damage from slight movement can be repaired once it has stabilised. Repoint the brickwork and apply mastic sealant to gaps round the woodwork. Damage from extensive or persistent movement should be dealt with by a builder. Consult your local Building Control Officer and inform your house insurance company.

Fitting the frame

Where the height of the original window permits it, standard window frames can be used to make up a replacement bay window. Using gasket seals (◁), various combinations of frames can be arranged. Shaped hardwood corner posts are available to give a 90, 60 or 45 degree angle to the side frames. The gasket is used for providing a weatherproof seal between the posts and the frames.

Joining frames
A flexible gasket, which is sold by the metre, is available for joining standard frames. The frames are screwed together to fit the opening.

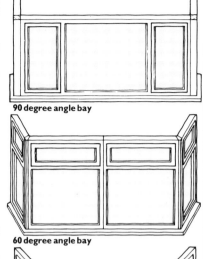
90 degree angle bay

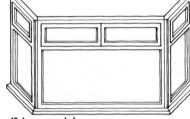
60 degree angle bay

45 degree angle bay

Hardwood bay window (Detail)

Bow windows

These are windows constructed on a shallow curve, and they normally project from a flat wall. Complete hardwood bow window frames are available from joinery suppliers, ready for installation in a brickwork opening. A flat-topped canopy of moulded plastic is made for finishing the top of the window. Bow windows can be substituted for conventional ones.

Fitting the frame

Tack a damp-proof course material to the sides of the frame and the underside of the sill, then fit the frame and the canopy into the wall opening together, the outer edges of the frame set flush with the wall finish. Screw the frame to the brickwork. The vertical damp-proofing should overlap the one in the wall if one is present.

Weatherproof the canopy with a lead flashing cut into the wall and dressed over the canopy upstand (◁). Use mastic to seal the joints between the frame's sides and sill and the brickwork.

Bow window (Detail) with leaded lights

REPLACING A SASH WINDOW

An old vertically sliding sash window with cords and counterweights can be replaced with a new frame fitted with spiral balance sashes.

Remove the sashes, then take out the old frame from inside the room. Prise off the architrave, then the window boards, and chop away the plaster as necessary. Most frames make a wedge fit, and you can loosen one by hitting the sill on the outside with a heavy hammer and a wood block. Lift the frame out of the opening when it is loose (**1**) and remove any debris from the opening once it is clear.

Set the new frame centrally in the opening so that its stiles are showing equal amounts on each side of the exterior brickwork reveals. Check the frame for plumb and wedge the corners at the head and the sill. Make up the space left by the old box stiles with mortared brickwork (**2**).

Metal brackets screwed to the frame's stiles can also be set in the mortar bed joints to secure the frame.

When the mortar is set, replaster the inner wall and replace the architrave. In the meantime glaze the sashes. Finally apply a mastic sealant to the joints between the outside brickwork and the frame to keep the weather out.

1 Lift out old frame

2 Fill gaps with brick

TYPES OF GLASS

Glass is made from silica sand which, with such additives as soda, lime and magnesia, is heated until it is molten to produce the raw material. The type and quality of the glass produced for windows is determined by the method by which it is processed in the molten stage. Ordinary window glass is known as annealed glass. Special treatments during manufacture give glass particular properties, such as heat-resistance or extra strength.

SHEET GLASS

Clear sheet glass is made for general glazing and was once the most common type used in windows. It is produced by a drawn process and can sometimes be recognised by its slight distorting effect, which is a result of the manufacturing process. Though the surfaces are given a smooth 'fire-finish' they are not always quite flat or quite parallel.

Two qualities of clear glass are produced – the ordinary standard grade for general glazing and a 'selected' one for better glazing work – in thicknesses from 2mm to 4mm ($^1/_{16}$ to $^5/_{32}$in).

Horticultural glass is a poorer category produced by the same method and made in 3mm ($^1/_8$in) and 4mm ($^5/_{32}$in) thicknesses for use in greenhouses only.

FLOAT GLASS

Float glass is now generally used for glazing windows. It is made by floating the molten glass on a bath of liquid tin to produce a sheet with flat, parallel and distortion-free surfaces. It has virtually replaced plate glass, which was a rolled glass ground and polished on both sides.

Clear float glass is made in thicknesses from 3mm ($^1/_8$in) up to 25mm (1in), but it is generally stocked in only three: 3mm ($^1/_8$in), 4mm ($^5/_{32}$in) and 6mm ($^1/_4$in).

PATTERNED GLASS

Patterned glass is glass which has one surface embossed with a texture or a decorative design. It is available in clear or tinted sheets in thicknesses of 3mm ($^1/_8$in), 4mm ($^5/_{32}$in), 5mm ($^3/_{16}$in) and 6mm ($^1/_4$in).

The transparency of the glass depends on the density of the patterning. Such 'obscured' glass is used where maximum light is required while maintaining privacy, such as in bathroom windows.

Textured glass, commonly known as rough-cast glass, is another form of patterned glass. It is made in a thicker range than patterned glass, 6mm ($^1/_4$in) being the most common thickness. It is used mostly for commercial buildings.

SOLAR CONTROL GLASS

There is now special glass available which cuts down the heat received from the sun, and this is often used in roof lights. This tinted glass, which can be of the float, sheet, laminated or rough-cast type, also reduces glare, though at the expense of some illumination. It is available in thicknesses ranging from 3mm ($^1/_8$in) to 12mm ($^1/_2$in), depending on the type.

NON-REFLECTIVE GLASS

Diffuse reflection glass is a 2mm ($^1/_{16}$in) glass sheet with slightly textured surfaces, and is much used for glazing picture frames. When placed within 12mm ($^1/_2$in) of the picture surface the glass appears completely transparent while eliminating the surface reflections associated with ordinary polished glass.

SAFETY GLASS

Glass which has been strengthened by means of reinforcement or a toughening process is known as safety glass. It should be used whenever the glazed area is relatively large or where its position makes it vulnerable to accidental breakage. Such hazardous areas common in domestic situations are glazed doors, low-level windows and shower screens.

WIRED GLASS

Wired glass is a 6mm ($^1/_4$in) thick rough-cast or clear annealed glass with a fine steel wire mesh incorporated in it during its manufacture. Glass with a 13mm ($^1/_2$in) square mesh is known as Georgian wired glass. The mesh supports the glass and prevent it disintegrating in the event of breakage. The glass itself is not special and is no stronger than ordinary glass of the same thickness.

Wired glass is regarded as a fire-resistant material with a one-hour rating. Though the glass may break, its wire reinforcement helps to maintain its integrity and prevent the spread of fire.

TOUGHENED GLASS

Toughened glass is ordinary glass that has been heat-treated to improve its strength, and is sometimes referred to as tempered glass. The process of treatment renders the glass about four to five times stronger than an untreated glass of similar thickness. In the event of its breaking it merely shatters into relatively harmless granules.

Toughened glass cannot be cut. Any work required, such as holes to be drilled for screws, must be done before the toughening process. Joinery suppliers of doors and windows usually stock standard sizes of toughened glass to fit standard frames.

LAMINATED GLASS

Laminated glass is made by bonding together two or more layers of glass with a clear tear-resistant plastic film sandwiched between. This safety glass will absorb the energy of an object hitting it, so preventing the object penetrating the pane. The plastic interlayer also binds broken glass fragments together and reduces the risk of injury from fragments of flying glass.

It is made in a range of thicknesses from 4mm ($^5/_{32}$in) up to 8mm ($^5/_{16}$in), depending on the type of glass used. Clear, tinted and patterned versions are all available.

BUYING GLASS

You can buy most types of glass from your local stockists. They will advise you on thickness, will cut the glass to your measurements and will also deliver larger sizes and amounts.

The thickness of glass, once expressed by weight, is now measured in millimetres. If you are replacing old glass, measure its thickness to the nearest millimetre, and, if it is slightly less than any available size, buy the next one up for the sake of safety.

Though there are no regulations about the thickness of glass, for safety reasons you should comply with the recommendations set out in the British Standard Code of Practice. The required thickness of glass depends on the area of the pane, its exposure to wind pressure and the vulnerability of its situation – e.g., in a window overlooking a play area. Tell your supplier what the glass is needed for – a door, a window, a shower screen etc. – to ensure that you get the right type.

Measuring up

Measure the height and width of the opening to the inside of the frame rebate, taking the measurement from two points for each dimension. Also check that the diagonals are the same length. If they differ markedly and show that the frame is out of square, or if it is otherwise awkwardly shaped, make a cardboard template of it. In any case deduct 3mm (⅛in) from the height and width to allow a fitting tolerance. When making a template allow for the thickness of the glass cutter.

When you order patterned glass, specify the height before the width. This will ensure that the glass is cut with the pattern running in the right direction. (Alternatively take a piece of the old glass with you, which you may need to do in any case to match the pattern.)

For an asymmetrically shaped pane of patterned glass supply a template, and mark the surface that represents the outside face of the pane. This ensures that the glass will be cut with its smooth surface outside and will be easier to keep clean.

WORKING WITH GLASS

You should always carry glass on its edge. You can hold it with pads of folded rag or paper when gripping the top and bottom edge, though it is always better to wear stout work gloves.

Protect your hands with gloves and your eyes with goggles when removing broken glass from a frame. Wrap up the broken pieces in thick layers of newspaper if you have to dispose of it in your dustbin, but before doing so check with your local glazier, who may be willing to take the pieces from you and add them to his off-cuts, which are usually sent back to the manufacturers for recycling.

Basic glass-cutting

It is usually unnecessary to cut one's own glass as glass merchants are willing to do it, but you may have some surplus glass and wish to cut it yourself. Diamond-tipped cutters are available, but the type with a steel wheel is cheaper and quite adequate for normal use.

Cutting glass successfully is largely a matter of practice and confidence. If you have not done it before, you should make a few practice cuts on waste pieces of glass and get used to the 'feel' before doing a real job.

Lay the glass on a flat surface covered with a blanket. Patterned glass is placed patterned side downwards and cut on its smooth side. Clean the surface with methylated spirit.

Set a T-square the required distance from one edge, using a steel measuring tape (**1**). If you are working on a small piece of glass or do not have a T-square, mark the glass on opposing edges with a felt-tipped pen or wax pencil and use a straight edge to join up the marks and guide the cutter.

Lubricate the steel wheel of the glass cutter by dipping it in thin oil or paraffin. Hold the cutter between middle finger and forefinger (**2**) and draw it along the guide in one continuous stroke. Use even pressure throughout and run the cut off the end. Slide the glass forward over the edge of the table (**3**) and tap the underside of the scored line with the back of the cutter to initiate the cut. Grip the glass on each side of the score line with gloved hands (**4**), lift the glass and snap it in two. Alternatively, place a pencil under each end of the scored line and apply even pressure on both sides until the glass snaps.

1 Measure the glass with a tape and T-square

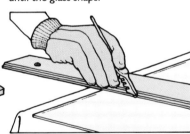

2 Cut glass in one continuous stroke

3 Tap the edge of glass to initiate the cut

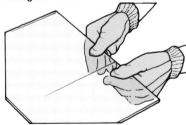

4 Snap glass in two

Cutting a thin strip of glass

A pane of glass may be slightly oversize due to inaccurate measuring or cutting or if the frame is distorted.

Remove a very thin strip of glass with the aid of a pair of pliers. Nibble away the edge by gripping the waste with the tip of the jaws close to the scored line.

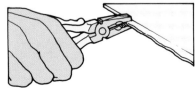

Nibble away a thin strip with pliers

CUTTING CIRCLES AND DRILLING HOLES

Fitting items such as an extractor fan may involve cutting a circular hole in a pane of glass. This can be done with a beam compass glass cutter.

Cutting a circle in glass

Locate the suction pad of the central pivot on the glass, set the cutting head at the required distance from it and score the circle round the pivot with even pressure (1). Now score another smaller circle inside the first one. Remove the cutter and score across the inner circle with straight cuts, then make radial cuts about 25mm (1in) apart in the outer rim. Tap the centre of the scored area from underneath to open up the cuts (2) and remove the inner area. Next tap the outer rim and nibble away the waste with pliers if necessary.

To cut a disc of glass, scribe a circle with the beam compass cutter, then score tangential lines from the circle to the edges of the glass (3). Tap the underside of each cut, starting close to the edge of the glass.

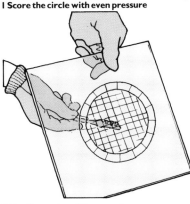

1 Score the circle with even pressure

SEE ALSO

Details for: ▷
Removing glass   24

Smoothing the edges of cut glass

You can grind down the cut edges of glass to a smooth finish using wet-and-dry paper wrapped round a wooden block. It is fairly slow work, though just how slow will depend on the degree of finish you require.

Start off with medium-grit paper wrapped tightly round the wood block. Dip the block complete with paper in water and begin by removing the 'arris' or sharp angle of the edge with the block held at 45 degrees to the edge. Keep the abrasive paper wet.

Follow this by rubbing down the vertical edge to remove any nibs and go on to smooth it to a uniform finish. Repeat the process with progressively finer grit papers. A final polish can be given with a wet wooden block coated with pumice powder.

2 Tap the centre of the scored area

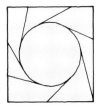

3 Cutting a disc
Scribe the circle then make tangential cuts from it to the edge of the glass.

Using a glass cutting template

Semi-circular windows and glazed openings in Georgian-style doors are formed with segments of glass set between radiating glazing-bars.

Windows with semi-circular openings and modern reproductions of period doors can be glazed with ready-shaped panes available from joinery suppliers, but for an old glazed door you will probably have to cut your own. The pieces are segments of a large circle, beyond the scope of the beam compass glass cutter (See above), so you will have to make a card template.

Remove the broken glass, clean up the rebate, then tape a sheet of paper over the opening and, using a wax crayon, take a rubbing of the shape (1). Remove the paper pattern and tape it to a sheet of thick cardboard. Following the lines on the paper pattern, cut the card to shape with a sharp knife, but make the template about 2mm (1/16in) smaller all round, also allowing for the thickness of the glass cutter. The straight cuts can be aided by a straightedge, but you will have to make curved ones freehand. A slightly wavy line will be hidden by the frame's rebate.

Fix the template to the glass with double-sided tape, score round it with the glass cutter (2), running all cuts to the edge, and snap the glass in the normal way.

1 Take a rubbing of the shape with a crayon

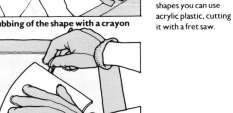

2 Cut round the template with even pressure

● **Plastic glazing**
As an alternative to glass for awkward shapes you can use acrylic plastic, cutting it with a fret saw.

Drilling a hole in glass

There are special spear-point drilling bits available for drilling holes in glass. As glass should not be drilled at high speed, use a hand-held wheel brace.

Mark the position for the hole, no closer than 25mm (1in) to the edge of the glass, using a felt-tipped pen or a wax pencil. On mirror glass work from the back, or coated surface.

Place the tip of the bit on the marked centre and, with light pressure, twist it back and forth so that it grinds a small pit and no longer slides off the centre. Form a small ring with putty round the pit and fill the inner well with a lubricant such as white spirit, paraffin or water.

Work the drill at a steady speed and with even pressure. Too much pressure can chip the glass.

When the tip of the drill just breaks through, turn the glass over and drill from the other side. If you try to drill straight through from one side you risk breaking out the surface round the hole.

Drilling glass
Always run the drill in a lubricant to reduce friction.

REPAIRING A BROKEN WINDOW

A cracked window pane, even when no glass is missing from it, is a safety hazard and a security risk. If the window is actually lacking some of its glass, it is no longer weatherproof and should be repaired promptly.

PAGE 202

SEE ALSO

◁ Details for:
Scaffold tower   78
Buying glass   22

Temporary repairs

For temporary protection from the weather a sheet of polythene can be taped or pinned with battens over the outside of the window frame, and a merely cracked window can be temporarily repaired with a special clear self-adhesive waterproof tape. Applied to the outside, the tape gives an almost invisible repair.

Safety with glass

The method you use to remove the glass from a broken window will to some extent depend on conditions. If the window is not at ground level, it may be safest to take out the complete sash to do the job. But a fixed window will have to be repaired on the spot, wherever it is.

Large pieces of glass should be handled by two people and the work done from a tower rather than ladders (◁). Avoid working in windy weather and always wear protective gloves for this work.

Repairing glass in wooden frames

In wooden window frames the glass is set into a rebate cut in the frame's moulding and bedded in linseed oil putty. Small wedge-shaped nails known as sprigs are also used to hold the glass in place. In some wooden-framed windows a screwed-on beading is used to hold the pane instead of the 'weathered' (outer) putty; this type of frame may have its rebate cut on the inside instead of the outside.

Removing the glass

If the glass in a window pane has shattered, leaving jagged pieces set in the putty, grip each piece separately (wearing gloves) and try to work it loose (1). It is safest always to start working from the top of the frame.

Old dry putty will usually give way, but if it is strong it will have to be cut away with a glazier's hacking knife and a hammer (2). Alternatively, the job can be done with a blunt wood chisel. Work along the rebate to remove the putty and glass. Pull out the sprigs with pincers (3).

If the glass is cracked but not holed, run a glass cutter round the perimeter of the pane about 25mm (1in) from the frame, scoring the glass (4). Fasten strips of self-adhesive tape across the cracks and the scored lines (5) and tap each piece of glass so that it breaks free and is held only by the tape. Carefully peel the inner pieces away, then remove the pieces round the edges and the putty as described above.

Clean out the rebate and seal it with a wood primer. Measure the height and width of the opening to the inside of the rebates and have your new glass cut 3mm (⅛in) smaller on each dimension (◁) to give a fitting tolerance.

Fitting new glass

Purchase new sprigs and enough putty for the frame. Your glass supplier should be able to advise you on this but, as a guide, 500g (1lb) of putty will fill an average-sized rebate of about 4m (13ft) in length.

Knead a palm-sized ball of putty to an even consistency. Very sticky putty is difficult to work with so wrap it briefly in newspaper to absorb some of the oil. You can soften putty that is too stiff by adding linseed oil to it.

Press a fairly thin, continuous band of putty into the rebate all round with your thumb. This is the bedding putty. Lower the edge of the new pane on to the bottom rebate, then press it into the putty. Press close to the edges only, squeezing the putty to leave a bed about 2mm (1/16in) behind the glass, then secure the glass with sprigs about 200mm (8in) apart. Tap them into the frame with the edge of a firmer chisel so that they lie flat with the surface of the glass (1). Trim the surplus putty from the back of the glass with a putty knife.

Apply more putty to the rebate all round, outside the glass. With a putty knife (2), work the putty to a smooth finish at an angle of 45 degrees. Wet the knife with water to prevent it dragging and make neat mitres in the putty at the corners. Let the putty set and stiffen for about three weeks, then apply an oil-based undercoat paint. Before painting, clean any putty smears from the glass with methylated spirit. Let the paint lap the glass slightly to form a weather seal.

A self-adhesive plastic foam can be used instead of the bedding putty. Run it round the back of the rebate in a continuous strip, starting from a top corner, press the glass into place on the foam and secure it with sprigs. Then apply the weathered putty in the same way described above.

Alternatively, apply a strip of foam round the outside of the glass and cover it with a wooden beading, then paint.

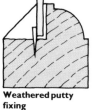

Weathered putty fixing

Wooden bead fixing
Unscrew beading and scrape out mastic. Bed new glass in fresh mastic and replace beading.

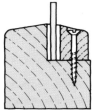

1 Work loose the broken glass

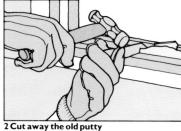

2 Cut away the old putty

3 Pull out the old sprigs

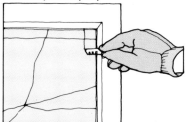

4 Score glass before removing a cracked pane

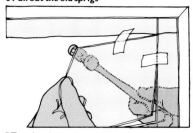

5 Tap the glass to break it free

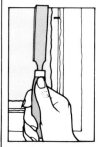

1 Tap in new sprigs

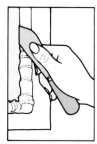

2 Shape the putty

DOOR REPAIRS

The great majority of external door frames are constructed of softwood, and this, if it is regularly maintained with a good paint system (▷), will give years of excellent service. However, the ends of door sills and the frame posts are vulnerable to wet rot if they are subject to continual wetting. This can happen when the frame has moved because of shrinkage of the timber, or where old pointing has fallen out and left a gap where water can get in. Alternatively, old and porous brickwork or an ineffective damp-proof course (▷) can be the cause of wet rot damage.

Prevention is always better than any cure, so check round the frame for any shrinkage gaps and apply a mastic sealant where necessary. Keep all pointing in good order. A slight outbreak of rot can be treated with the aid of a proprietary repair kit and preservative (▷).

Replacing a sill

You can buy 150 x 50mm (6 x 2in) softwood or hardwood door sill sections which can be cut to the required length. If your sill is not of a standard-shaped section the replacement can be made to order. It is more economical in the longer run to specify a hardwood such as oak or utile, as it will last much longer.

Taking out the old
First measure and note down the width of the door opening, then remove the door. The posts are usually tenoned into the sill, so to separate the sill from them split it lengthwise with a wood chisel. A saw cut across the centre of the sill can make the job easier.

The ends of the sill are set into the brickwork on either side, so cut away the bricks to make the removal of the old sill and insertion of the new one easier. Use a plugging chisel (▷) to cut carefully through the mortar round the bricks and try to preserve them for reuse after fitting the sill.

The new sill has to be inserted from the front so that it can be tucked under the posts and into the brickwork. Cut the tenons off level with the shoulders of the posts (1). Mark and cut shallow housings for the ends of the posts in the top of the new sill, spaced apart as previously noted. The housings must be deep enough to take the full width of the posts (2), which may mean the sill being slightly higher than the original one, so that you will have to trim a little off the bottom of the door.

Fitting the new
Try the new sill for fit and check that it is level. Before fixing it apply a wood preservative to its underside and ends,

It is possible for the sill to rot without the frame posts being affected. In this case just replace the sill. If the posts are also affected, repair them (See right). In some cases the post ends can be tenoned into the sill and fitted as a unit.

and, as a measure against rising or penetrating damp, apply two or three coats of bitumen latex emulsion to the brickwork (▷).

When both treatments are dry, glue the sill to the posts, using an exterior woodworking adhesive. Wedge the underside of the sill with slate to push it up against the ends of the posts, skew-nail the posts to it and leave it for the adhesive to set.

Pack the gap between the underside of the sill and the masonry with a stiff mortar of 3 parts sand: 1 part cement, and rebond and point the bricks. Finish by treating the wood with preservative and applying a mastic sealant round the door frame.

1 Cut tenons off level with the joint's shoulder

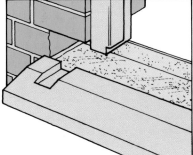

2 Cut a housing to receive the post

REPAIRING DOOR POSTS

Rot can attack the ends of door posts, particularly in exposed positions where they meet stone steps or are set into concrete, as is found in some garages. The posts may be located on metal dowels set into the step.

If the damage is not too extensive the rotten end can be cut away and replaced with a new piece, either scarf jointed or half lap jointed into place. If your situation involves a wooden sill combine the following information with that given for replacing a sill (See left).

First remove the door, then saw off the end of the affected post back to sound timber. For a scarf joint make the cut at 45 degrees to the face of the post (1). For a lap joint cut it square. Chip any metal dowel out of the step with a cold chisel.

Measure and cut a matching section of post to the required length, allowing for the overlap of the joint, then cut the end to 45 degrees or mark and cut a half lap joint in both parts of the post (2).

Drill a hole in the end of the new section for the metal dowel if it is still usable. If it is not, make a new one from a piece of galvanized steel gas pipe, priming the metal to prevent corrosion Treat the new timber with a preservative and insert the dowel. Set the dowel in mortar, at the same time gluing and screwing the joint (3).

If a dowel is not used, fix the post to the wall with counterbored screws. Place hardboard or plywood packing behind it if necessary and plug the counterbores of the screw holes.

Apply a mastic sealant to the joints between the door post, wall and base.

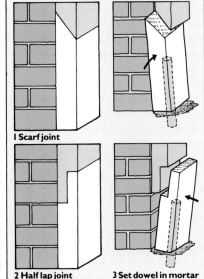

1 Scarf joint

2 Half lap joint   **3 Set dowel in mortar**

25

FITTING AND HANGING DOORS

Whatever the style of door you wish to fit, the procedure is the same, though minor differences between some external doors may show themselves. Two good-quality 100mm (4in) butt hinges are enough to support a standard door, but if you are hanging a heavy hardwood one you should add a third, central hinge.

All doors are fairly heavy, and as it is necessary to try a door in its frame several times to get the fit right you will find that the job goes much more quickly and easily if you have a helper working with you.

Fitting a door

Before attaching the hinges to a new door make sure that it fits nicely into its frame. It should have a clearance of 2mm (1/16in) at the top and sides and should clear the floor by at least 6mm (1/4in). As much as 12mm (1/2in) may be required for a carpeted floor.

Measure the height and width of the door opening and the depth of the rebate in the door frame into which the door must fit. Choose a door of the right thickness and, if you cannot get one that will fit the opening exactly, one which is large enough to be cut down.

Cutting to size

Some doors are supplied with 'horns', extensions to their stiles which protect the corners while the doors are in storage. Cut these off with a saw (**1**) before starting to trim the door to size.

Transfer the measurements from the frame to the door, making necessary allowance for the clearances all round. To reduce the width of the door stand it on edge with its latch stile upwards while it is steadied in a portable vice. Plane the stile down to the marked line, working only on the one side if a small amount is to be taken off. If a lot is to be removed, take some off each side. This is especially important with panel doors to preserve the symmetry.

If you need to take off more than 6mm (1/4in) to reduce the height of the door, remove it with a saw and finish off with a plane. Otherwise plane the waste off (**2**). The plane must be sharp to deal with the end grain of the stiles. Work from each corner towards the centre to avoid 'chipping out' the corners.

Try the door in the frame, supporting it on shallow wedges (**3**). If it still doesn't fit take it down and remove more wood where appropriate.

1 Saw off horns

2 Plane to size

3 Wedge the door

Fitting hinges

The upper hinge is set about 175mm (7in) from the door's top edge and the lower one about 250mm (10in) from the bottom. They are cut equally into the stile and door frame. Wedge the door in its opening and, with the wedges tapped in to raise it to the right floor clearance, mark the positions of the hinges on both the door and frame.

Stand the door on edge, the hinge stile uppermost, open a hinge and, with its knuckle projecting from the edge of the door, align it with the marks and draw round the flap with a pencil (**1**). Set a marking gauge (◁) to the thickness of the flap and mark the depth of the housing. With a chisel make a series of shallow cuts across the grain (**2**) and pare out the waste to the scored line. Repeat the procedure with the second hinge, then, using the flaps as guides, drill pilot holes for the screws and fix both hinges into their housings.

Wedge the door in the open position, aligning the free hinge flaps with the marks on the door frame. Make sure that the knuckles of the hinges are parallel with the frame, then trace the housings on the frame (**3**) and cut them out as you did the others.

Adjusting and aligning

Hang the door with one screw holding each hinge and see if it closes smoothly. If the latch stile rubs on the frame you may have to make one or both housings slightly deeper. If the door strains against the hinges it is what is called 'hinge bound'. In this case insert thin cardboard beneath the hinge flaps to pack them out. When the door finally opens and closes properly drive in the rest of the screws.

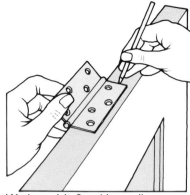

1 Mark round the flap with a pencil

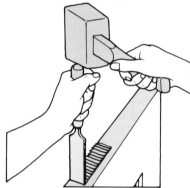

2 Cut across the grain with a chisel

3 Mark the size of the flap on the frame

MEASUREMENTS

A door that fits well will open and close freely and look symmetrical in the frame. Use the figures given as a guide for trimming the door and setting out the position of the hinges.

3mm (1/8in) clearance at top and sides ●
Upper hinge 175mm (7in) from the top ●

Lower hinge 250mm (10in) from the bottom ●
6 to 12mm (1/4 to 1/2in) gap at the bottom ●

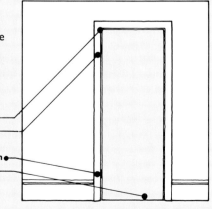

Rising butt hinges

Rising butt hinges lift a door as it is opened and are fitted to prevent it dragging on thick pile carpet.

They are made in two parts: a flap, with a fixed pin, which is screwed to the door frame, and another, with a single knuckle, which is fixed to the door, the knuckle sliding over the pin.

Rising butt hinges can be fixed only one way up, and are therefore made specifically for left- or right-hand opening. The countersunk screwholes in the fixed pin flap indicate the side to which it is made to be fitted.

Fitting

Trim the door and mark the hinge positions (See opposite), but before fitting the hinges plane a shallow bevel at the top outer corner of the hinge stile so that it will clear the frame as it opens. As the stile (▷) runs through to the top of the door, plane from the outer corner towards the centre to avoid splitting the wood. The top strip of the door stop will mask the bevel when the door is closed.

Fit the hinges to the door and the frame, then lower the door on to the hinge pins, taking care not to damage the architrave above the opening.

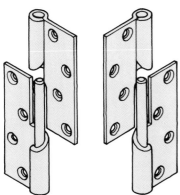

Left-hand opening   **Right-hand opening**

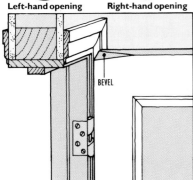

BEVEL

Plane a shallow bevel to clear the door frame

Weatherproofing a door

Fitting a weatherboard

A weatherboard is a special moulding fitted to the bottom of an outer door to shed rainwater away from the threshold. To fit one measure the width of the opening between the door stops and cut the moulding to fit, cutting one end at a slight angle where it meets the door frame on the latch side. This will allow it to clear the frame as the door swings open.

Make a weatherproof seal between the moulding and the door. On an unfinished door use screws and a waterproof adhesive to attach the moulding. On a pre-painted one apply a thick coat of primer to the back surface of the moulding and screw it into place while the primer is still wet. Fill or plug all screwholes and thoroughly prime and finish the surfaces.

Allowing for a weather bar

Though a rebate cut into the head and side posts of an outer door frame provides a seal round an inward opening door, a rebate cut into the sill at the foot of the door would merely encourage water to flow into the house.

Unless protected by a porch, a door in an exposed position needs to be fitted with a weather bar to prevent rainwater running underneath.

This is a metal or plastic strip which is set into the step or sill. If you are putting in a new door and wish to fit a weather bar, use a router or power saw to cut a rebate across the bottom of the door in order to clear the bar.

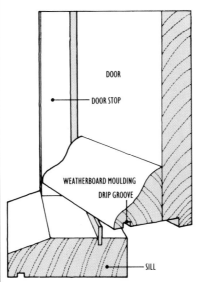

DOOR

DOOR STOP

WEATHERBOARD MOULDING
DRIP GROOVE

SILL

Door fitted with a weather board

ADJUSTING BUTT HINGES

Perhaps you have a door catching on a bump in the floor as it opens. You can, of course, fit rising butt hinges, but the problem can be overcome by resetting the lower hinge so that its knuckle projects slightly more than the top one. The door will still hang vertically when closed, but as it opens the out-of-line pins will throw it upwards so that the bottom edge will clear the bump.

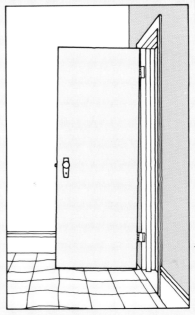

Resetting the hinge
You may have to reset both hinges to the new angle to prevent binding.

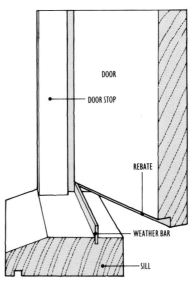

DOOR

DOOR STOP

REBATE

WEATHER BAR

SILL

Sill fitted with weather bar

SEE ALSO

Details for: ▷
Stile   78

REPAIRS AND IMPROVEMENTS

Repairing a battened door

The battens, or tongue-and-groove jointed boards, of a ledged and braced cottage or garage door can rot at the bottom because of the end grain absorbing moisture. The damage can be patch-repaired. Nailing a board across the bottom of the door is not the easy solution it may appear because moisture will be trapped behind the board and will increase the rot.

Remove the door and cut back the damaged boards to sound material. Where a batten falls on a rail, use the tip of a tenon saw, held at a shallow angle, to cut through most of it, then finish off the cut with a chisel. Use a padsaw or a power jigsaw where the blade can pass clear of the rail.

When replacing the end of a single batten make the cut at right angles (**1**). When a group of battens is to be replaced make 45 degree cuts across them (**2**). In this way the interlocking of the tongued and grooved edges between old and new sections is better maintained.

When cutting new pieces of boarding to fit, leave them over length. Apply an exterior woodworking adhesive to the butting ends of the battens but take care not to get any on the tongue-and-groove joints. Tap the pieces into place and nail each to the rail with two staggered lost head nails. Cut off the ends in line with the door's bottom edge and treat the wood with a good preservative (◁) to prevent any further rot damage.

SEE ALSO
◁ Details for:
Wood preservative   16

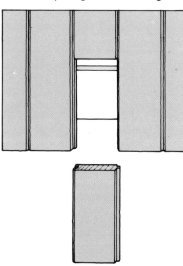

1 Cut the end of a single batten square

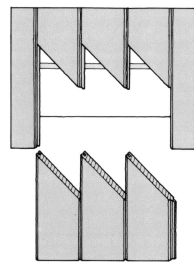

2 Cut a group of battens to 45 degrees

Easing a sticking door

If the bottom corner of a door is rubbing on the frame it is probably swollen. Take it off its hinges and shave the corner down with a plane. If the top corner is rubbing check the hinges before doing any planing. After years of use, hinges become worn so that the pins are slack and the door drops. In this case fit new hinges, or, as a cheaper alternative, swap the old hinges, the top with the bottom, which will reverse the wear on the pins.

Swopping hinges ▶
Swop worn hinges, top with bottom, for a cheap and convenient repair.

FLUSH-PANELLING A DOOR

An old framed and panelled door can be given a new lease of life by flush-panelling it with a sheet of hardboard or wood-veneered panelling on each side.

Using hardboard
The easiest way is to cut and fit a hardboard panel 25mm (1in) smaller all round than the door. This can be done without removing the door.

Sandpaper the surface of the frame to provide a key, then apply contact adhesive to the frame members and to the corresponding areas on the back of the hardboard. Position the panels carefully and fix the edges with hardboard nails at 150mm (6in) intervals. Tap all over the glued areas with a wood block and a hammer.

Fill the nail recesses, prime the surfaces and paint as required.

Using a wood-veneered panel
This should extend to all four edges of the door, so the door must first be removed from the frame and the hinges and latch taken off.

Remove any paint from the long edges, scrape them back to bare wood and sandpaper them smooth.

Cut the panels slightly larger than the door, key the surface of the frame with abrasive paper and bond the panels to it with a contact adhesive.

When the adhesive is set plane the panel edges flush with the door all round. With a chisel trim away the edge of the panel where it overlaps the hinge housing. Stain the stripped edges of the door and edges of the panel to match the colour of the veneer.

Before rehanging the door remove the door stop strips. Fit the hinges and latch, rehang the door and replace the door stops against the door while it is in the closed position.

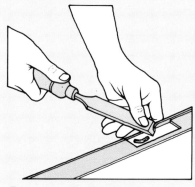

Trim the panelling at the hinge housings.

REPAIRING CONCRETE

Concrete is used in and around the house as a surface for solid floors, drives, paths and walls. In common with other building materials, it suffers from the effects of damp – spalling and efflorescence – and related defects such as cracking and crumbling. Repairs can usually be made in much the same way as for brickwork and render, although there are some special considerations you should be aware of. If the damage is widespread, however, it's quite straightforward to resurface it prior to decorating.

Sealing concrete

New concrete has a high alkali content and efflorescence can develop on the surface as it dries out. Do not use any finish other than a water-thinned paint until the concrete is completely dry. Treat efflorescence on concrete as for brickwork (▷).

A porous concrete wall should be water-proofed with a clear sealant on the exterior. Some reinforced emulsions will cover bitumen satisfactorily but it will bleed through most paints unless you prime it with a PVA bonding agent diluted 50 per cent with water. Alternatively, use an aluminium spirit-based sealer.

Cleaning dirty concrete

Clean dirty concrete as you would brickwork. Where a concrete drive or garage floor, for instance, is stained with patches of oil or grease, soak up fresh spillages immediately to prevent them becoming permanent stains. Sprinkle dry sand onto patches of oil to absorb any liquid deposits, collect it up and wash the area with white spirit or degreasing solution.

Binding dusty concrete

Concrete is trowelled when it is laid to give a flat finish; if this is overdone, cement is brought to the surface and when the concrete dries out, this thin layer begins to break up within a short time, producing a loose, dusty surface. You must not apply a decorative finish to concrete in this condition.

Treat a concrete wall with stabilizing primer but paint a dusty floor with one or two coats of PVA bonding agent mixed with five parts of water. Use the same solution to prime a particularly porous surface.

Making good cracks and holes

Rake out and brush away loose debris from cracks or holes in concrete. If the crack is less than about 6mm (¼in) wide, open it up a little with a cold chisel so that it will accept a filling. Undercut the edges to form a lip so the filler will grip.

To fill a hole in concrete, add a fine aggregate such as gravel to the sand and cement mix. Make sure the fresh concrete sticks in shallow depressions by priming the damaged surface with 3 parts bonding agent: 1 part water. When the primed surface becomes tacky, trowel in the concrete and smooth it.

Treating spalled concrete

When concrete breaks up or spalls due to the action of frost, the process is accelerated when steel reinforcement is exposed and begins to corrode. Fill the concrete as described above but prepare and prime the metalwork first If spalling recurs, particularly in exposed conditions, protect the wall with a bitumen base coat and a compatible reinforced emulsion paint.

Spalling concrete ▷
Rusting metalwork causes concrete to spall

DAMP-PROOFING A CONCRETE FLOOR

An uneven or pitted concrete floor must be made flat and level before you apply any form of floorcovering. You can do this fairly easily using a proprietary self-levelling screed. But first of all you must ensure the surface is free from dampness.

Testing for dampness
Do not lay any tiles or sheet floorcoverings (or apply a levelling screed) to a floor that's damp. A new floor should incorporate a damp-proof membrane (DPM) but must be left to dry out for six weeks before any covering is added.

If you suspect an existing floor is damp, make a simple test by laying a small piece of polythene on the concrete then seal it all round with adhesive tape. After one or two days, look to see if there are any traces of moisture on the underside.

For a more accurate assessment, hire a moisture meter. This device has contact pins or deep wall probes, which, when stuck into the suspect surface, gives a moisture saturation percentage: if the moisture reading does not exceed 6 per cent proceed with covering or levelling the floor.

If either test indicates treatment is necessary, paint the floor with a waterproofing compound. Prime the surface first with a slightly diluted coat, then brush on two full-strength coats, allowing each to dry between applications. If necessary you can then lay a self-levelling screed over the waterproofing compound.

Applying a self-levelling compound
Fill holes and cracks deeper than about 3mm (⅛in) by first raking out and undercutting the edges (1), then spreading self-levelling compound over the entire floor surface. This material is supplied as a powder, which you mix with water.

Make sure the floor is clean and free from damp, then pour some of the compound in a corner furthest from the door. Spread the compound with a trowel (2) until it is about 3mm (⅛in) thick, then leave it to seek its own level.

Continue across the floor, joining the area of compound until the entire surface is covered. You can walk on the floor after about one hour, but leave the compound to harden for a few days before laying permanent floorcovering.

SEE ALSO

Details for: ▷	
Efflorescence	32
Repairing brickwork	34
Repairing render	35
Masonry paints	41
Damp-proof membrane	30
Curing damp	7–14

1 Rake out cracks

2 Apply compound
Spread levelling compound with a trowel.

REPAIRING A CONCRETE FLOOR

Concrete floors can be subject to cracking caused by shrinkage. Usually the cracks are only in the screed and can be easily repaired, but a cracked floor that is also uneven may be a sign of settlement in the sub-base, and you should have it checked by a surveyor or by the Building Control Officer, who will advise you on what steps to take.

Filling a crack
Clean all dirt and loose material out of the crack and, if necessary, open up narrow parts with a cold chisel to allow better penetration of the filler.

Prime the crack with a solution of 1 part bonding agent : 5 parts water and let it dry. Make a filler of 3 parts sand : 1 part cement mixed with equal parts of bonding agent and water; or use a ready-mixed quick-setting cement. Apply the filler with a trowel, pressing it well into the crack.

Laying a pipe in a concrete floor
Installations like central heating call for pipework to run across a room. If the floor is solid that means either running it round the walls or setting it into the floor. The latter looks better, though it involves more work.

Cut a 100mm (4in) wide channel with a cold chisel and club hammer. It should be deep enough for the pipe to be set about 25mm (1in) below the surface while also having a clearance under it.

You may damage the damp-proof membrane while cutting the channel, in which case it must be repaired. Bitumen-based liquid damp-proofing can be used to restore a bitumen or polythene DPM. For a polythene one carefully cut back the screed about 100mm (4in) all round the channel to expose the polythene surface.

Remove all loose material and apply a primer coat of diluted damp-proofer according to the maker's instructions and allow it to dry.

With the pipework installed, apply two or three coats to the bottom and sides of the channel, stopping about 12mm (½in) from the surface. Allow each coat to dry before applying the next. The coatings must overlap the original DPM to form a continuous skin.

Test the pipework before filling the channel with a mix of 3 parts sand : 1 part cement and trowel it level.

LAYING A CONCRETE FLOOR

A suspended timber floor which has been seriously damaged by rot or insect infestation can be replaced with a solid concrete floor provided that the space below it needs no more than 600mm (24in) of infill material. (If this were the case a concrete floor would be liable to damage through settlement of the infill, so a new suspended floor would have to be fitted.)

Before taking any action consult your local Building Control Officer, because the converting of one floor can affect the ventilation of another.

If the work involves the electrical supply main or the supply pipes for gas or water you should check with the appropriate authority.

Wiring and heating pipes should be re-run before the infill is laid

Preparing the ground

Remove and burn all of the old infected timbers and take off the door of the room. Treat the ground and all the surrounding masonry thoroughly with a good fungicide (◁). With bricks and mortar fill in any recesses in the walls left by the timbers. Demolish the sleeper walls and combine the fragments with the infill material.

Mark the walls with a levelled chalk line to indicate the finished floor level, at the same time making allowance for the floor covering if you intend to use a thick material such as quarry tiles or wood blocks. About 50mm (2in) below this line mark another one, the space between them representing the thickness of the screed. Then mark a third line a further 100mm (4in) down, indicating the thickness of the slab.

The infill

Lay the infill material to the required depth in layers of no more than 225mm (9in) at a time, compacting each layer thoroughly and breaking up the larger pieces with a sledgehammer **(1)**. You can use brick and tile rubble or, better still, gravel rejects, which are coarse stones from quarry waste. If you are using secondhand rubble you should remove from it any fragments of plaster, which can react unfavourably with cement, and any pieces of wood.

Bring the surface of finely broken rubble up to within 25mm (1in) of the chalk line for the concrete and 'blind' the surface with a layer of sand, tamped or rolled flat.

The damp-proofing
Spread a sheet polythene damp-proof membrane of 1000-gauge (0.010in) minimum thickness over the surface of the sand turning its edges up all round and lapping it up the walls to form a tray. Make neat folds at the corners and hold them temporarily in place with paper clips. If the floor needs more than one sheet to cover it the sheets must overlap by at least 200mm (8in) and the joints be sealed with a special waterproof tape available from builders' merchants.

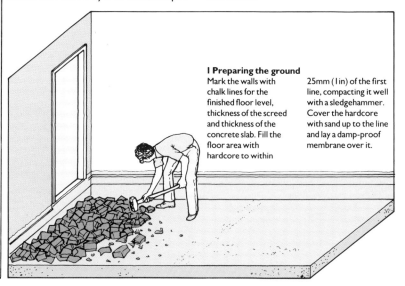

1 Preparing the ground
Mark the walls with chalk lines for the finished floor level, thickness of the screed and thickness of the concrete slab. Fill the floor area with hardcore to within 25mm (1in) of the first line, compacting it well with a sledgehammer. Cover the hardcore with sand up to the line and lay a damp-proof membrane over it.

Laying the concrete

Mix a medium-strength concrete of 1 part cement: 2½ parts sand: 4 parts aggregate. Do not add too much water; the mix should be a relatively stiff one.

Lay the concrete progressively in bands about 600mm (2ft) wide. The direction of the bands will depend on the door because you will have to work in such a way as to finish at the doorway. Tamp the concrete with a length of 100 x 50mm (4 x 2in) timber to compact it and finish level with the chalked line **(2)**. As you go along, check the overall surface with a spirit level and straight edge, and fill in any hollows, though slight unevennesses will be taken up by the screed.

When the concrete has set firmly enough to support a board to walk on, brush the surface with a stiff broom to make a key for the screed. Leave it to cure for at least three days under a sheet of polythene to prevent shrinkage caused by rapid drying.

Laying the screed

Mix the screed material from 3 parts sharp sand : 1 part portland cement. Dampen the floor and prime it with a cement grout mixed to a creamy consistency with water and bonding agent in equal parts. Working from one wall, apply a 600mm (2ft) band of grout with a stiff brush.

Apply a bedding of mortar at each end of the grouted area to take 38 x 38mm (1½ x 1½in) 'screed battens'. True them with a spirit level and straight edge so that they are flush with the surface-level line on the walls.

Lay mortar between the battens, tamp it well down **(3)** , level it with the straight edge laid across the battens and smooth it with a wooden float. Lift the battens out carefully, fill the hollows left with mortar and level with the float.

Repeat the procedure, working across the floor in bands 600mm (2ft) wide. Cover the finished floor with a sheet of polythene and leave it to cure for about a week.

The floor will be hard enough to walk on in two weeks, but not fully dry for about six months. Allow a month for every 25mm (1in) of thickness and meanwhile do not lay an impermeable floor covering.

Trim the damp-proof membrane to within 25mm (1in) of the floor and fit the skirtings to cover its edges **(4)**.

SEE ALSO

Details for: ▷
Repairing concrete 29–30

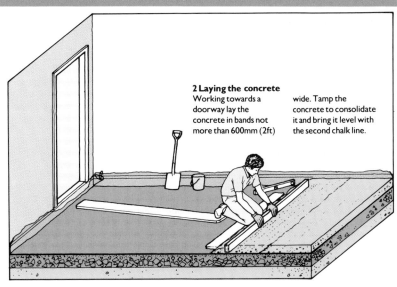

2 Laying the concrete
Working towards a doorway lay the concrete in bands not more than 600mm (2ft) wide. Tamp the concrete to consolidate it and bring it level with the second chalk line.

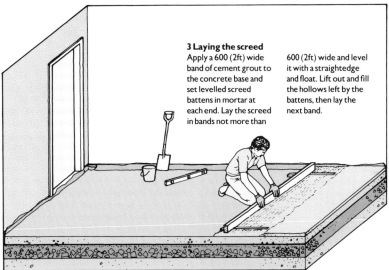

3 Laying the screed
Apply a 600 (2ft) wide band of cement grout to the concrete base and set levelled screed battens in mortar at each end. Lay the screed in bands not more than 600 (2ft) wide and level it with a straightedge and float. Lift out and fill the hollows left by the battens, then lay the next band.

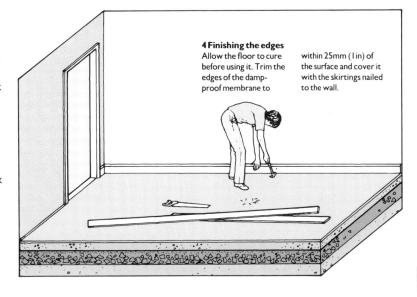

4 Finishing the edges
Allow the floor to cure before using it. Trim the edges of the damp-proof membrane to within 25mm (1in) of the surface and cover it with the skirtings nailed to the wall.

Before you decorate the outside of your house, check the condition of the brick and stonework, and carry out any necessary repairs. There's no reason why you can't paint brick or stone walls – indeed, in some areas it is traditional to do so – but if you consider masonry most attractive in its natural state, you could be faced with a problem: once masonry is painted, it is not possible to restore it to its original condition. There will always be particles of paint left in the texture of brickwork, and even smooth stone, which can be stripped successfully, may be stained by the paint.

Stained brickwork

Organic growth

Efflorescence

Treating new masonry

New brickwork or stonework should be left for about three months until it is completely dry before any further treatment is considered. White powdery deposits called efflorescence may come to the surface over this period, but you can simply brush it off with a stiff-bristled brush or a piece of dry sacking (◁). After that, bricks and mortar should be weatherproof and therefore require no further protection or treatment.

Cleaning organic growth from masonry

There are innumerable species of mould growth or lichens which appear as tiny coloured specks or patches on masonry. They gradually merge until the surface is covered with colours ranging from bright orange to yellow or green, grey and black.

Moulds and lichen will only flourish in damp conditions, so try to cure the source of the problem before treating the growth (◁). If one side of the house always faces away from the sun, it will have little chance to dry out. Relieve the situation by cutting back overhanging trees or shrubs to increase ventilation to the wall.

Make sure the damp-proof course (DPC) is working adequately and is not being bridged by piled earth or debris.

Cracked or corroded rainwater pipes leaking onto the wall are another common cause of organic growth. Feel behind the pipe with your fingers or use a hand mirror to locate the leak.

Removing the growth
Brush the wall vigorously with a stiff-bristled brush. This can be an unpleasant, dusty job, so wear a gauze facemask. Brush away from you to avoid debris being flicked into your eyes.

Microscopic spores will remain even after brushing. Kill these with a solution of bleach or, if the wall suffers persistently from fungal growth, use a proprietary fungicide, available from most DIY stores.

Using a bleach solution
Mix one part household bleach with four parts water. Paint the solution onto the wall using an old paintbrush, then 48 hours later wash the surface with clean water, using a scrubbing brush. Brush on a second application of bleach solution if the original fungal growth was severe.

Using a fungicidal solution
Dilute the fungicide with water according to the manufacturer's instructions and apply it liberally to the wall with an old paintbrush. Leave it for 24 hours then rinse the wall with clean water. In extreme cases, give the wall two washes of fungicide, allowing 24 hours between applications and a further 24 hours before washing it down with water.

Removing efflorescence from masonry

Soluble salts within building materials such as cement, brick, stone and plaster gradually migrate to the surface along with the water as a wall dries out. The result is a white crystalline deposit called efflorescence.

The same condition can occur on old masonry if it is subjected to more than average moisture. Efflorescence itself is not harmful but the source of the damp causing it must be identified and cured before decoration proceeds.

Regularly brush the deposit from the wall with a dry stiff-bristled brush or coarse sacking until the crystals cease to form – don't attempt to wash off the crystals; they'll merely dissolve in the water and soak back into the wall. Above all, don't attempt to decorate a wall which is still efflorescing, and therefore damp.

When the wall is completely dry, paint the surface with an alkali-resistant primer to neutralize the effect of the crystals before you paint with oil paint; water-thinned paints or clear sealants let the wall breathe, so are not affected by the alkali content of the masonry. Some specially formulated masonry paint can be used without primer (◁).

CLEANING OLD MASONRY

Whatever type of finish you intend to apply to a wall, all loose debris and dirt must be brushed off with a stiff-bristled brush. Don't use a wire brush unless the masonry is badly soiled as it may leave scratch marks.

Brush along the mortar joints to dislodge loose pointing. Defective mortar can be repaired easily at this stage (see right), but if you fail to disturb it now by being too cautious, it will fall out as you paint, creating far more work in the long run.

Removing unsightly stains

Improve the appearance of stone or brick left in its natural state by washing it with clean water. Play a hose gently onto the masonry while you scrub it with a stiff-bristled brush (1). Scrub heavy deposits with half a cup of ammonia added to a bucketful of water, then rinse again.

Abrade small cement stains or other marks from brickwork with a piece of similar-coloured brick, or scrub the area with a household kitchen cleanser.

Remove spilled oil paint from masonry with a proprietary paint stripper (▷). Put on gloves and protective goggles, then paint on the stripper, stippling it into the rough texture (2). After about ten minutes, remove it with a scraper and a soft wire brush. If paint remains in crevices, dip the brush in stripper and gently scrub it with small circular strokes. When the wall is clean, rinse with water.

1 Remove dirt and dust by washing

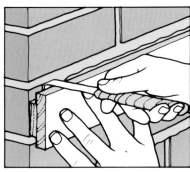
2 Stipple paint stripper onto spilled oil paint

REPOINTING MASONRY

The mortar joints between bricks and stone can become porous with age, allowing rainwater to penetrate to the inside, causing damp patches to appear, ruining decorations. Replacing the mortar pointing, which deflects the water, is quite straightforward but time consuming. Tackle only a small, manageable area at a time, using a ready-mixed mortar or your own mix.

Applying the pointing mortar

Rake out the old mortar pointing with a thin wooden lath to a depth of about 12mm (½in). Use a cold chisel or a special plugging chisel and a club hammer to dislodge firmly embedded sections, then brush out the joints with a stiff-bristled brush.

Flick water into the joints using an old paintbrush, making sure the bricks or stones are soaked so they will not absorb too much water from the fresh mortar. Mix up some mortar in a bucket and transfer it to a hawk. If you're mixing your own mortar, use the proportions 1 part cement: 1 part lime: 6 parts builders' sand.

Pick up a little sausage of mortar on the back of a small pointing trowel and push it firmly into the upright joints. This can be difficult to do without the mortar dropping off, so hold the hawk under each joint to catch it.

Try not to smear the face of the bricks with mortar, or it will stain. Repeat the process for the horizontal joints. The actual shape of the pointing is not vital at this stage.

Once the mortar is firm enough to retain a thumb print, it is ready for shaping. Match the style of pointing used on the rest of the house (see below). When the pointing has almost hardened, brush the wall to remove traces of surplus mortar.

Shaping the mortar joints

The joints shown here are commonly used for brickwork but they are also suitable for stonework. Additionally, stone may have raised mortar joints.

Flush joints

The easiest profile to produce, a flush joint is used where the wall is sheltered or painted. Rub each joint with sacking; start with the verticals.

Rubbed (rounded) joints

Bricklayers make a rubbed or rounded joint with a tool shaped like a sled runner with a handle: the semi-circular blade is run along the joints.

Improvise by bending a short length of metal tube or rod. Use the curved section only or you'll gouge the mortar. Alternatively, use a length of 9mm (⅜in) diameter plastic tube.

Raked joints

A raked joint is used to emphasize the type of bonding pattern of a brick wall. It's not suitable for soft bricks or for a wall that takes a lot of weathering. Scrape out a little of the mortar then tidy up the joints by running a 9mm (⅜in) lath along them.

Weatherstruck joints

The sloping profile is intended to shed rainwater from the wall. Shape the mortar with the edge of a pointing trowel. Start with the vertical joints, and slope them in either direction but be consistent. During the process, mortar will tend to spill from the bottom of a joint, as surplus is cut off. Bricklayers use a tool called a 'frenchman' to neaten the work: it has a narrow blade with the tip bent at right-angles. Make your own by bending a thin metal strip then bind insulating tape round the other end to form a handle, or bend the tip of an old kitchen knife after heating it in the flame of a blowtorch or cooker burner.

You will find it easiest to use a wooden batten to guide the blade of the frenchman along the joints, but nail scraps of plywood at each end of the batten to hold it off the wall.

Align the batten with the bottom of the horizontal joints, then draw the tool along it, cutting off the excess mortar, which drops to the ground.

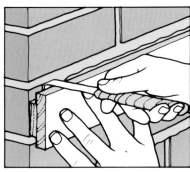
Use a frenchman to trim weatherstruck joints

SEE ALSO

Details for: ▷	
Paint stripper	78
Penetrating damp	7–8

● **Mortar dyes**
Liquid or powder additives are available for changing the colour of mortar to match existing pointing. Colour matching is difficult and smears can stain the bricks permanently.

Flush joint

Rubbed joint

Raked joint

Weatherstruck joint

33

REPAIRING MASONRY

REPAIRING SPALLED MASONRY

Cracks in external walls can be either the source of penetrating damp (◁), which ruins your decorations inside, or the result of a much more serious problem: subsidence in the foundations. Whatever the cause, it's obvious that you shouldn't just ignore the danger signs, but effect immediate cures.

Filling cracked masonry

If substantial cracks are apparent in a brick or stone wall, consult a builder or your local Building Control Officer to ascertain the cause

If the crack seems to be stable, it can be filled. Where the crack follows the line of the mortar joints, rake out those affected and repoint in the normal way, as previously described. A crack that splits one or more bricks or stones cannot be repaired, and the damaged area should be removed and replaced, unless you are painting the wall.

Use a ready-mixed mortar with a little PVA bonding agent added to help it to stick. Soak the cracked masonry with a hose to encourage the mortar to flow deeply into the crack.

Crack may follow pointing only

Cracked bricks could signify serious faults

Moisture penetrating soft masonry will expand in icy weather, flaking off the outer face of brickwork and stonework. The process, known as spalling, not only looks unattractive but also allows water to seep into the surface. Repairs to spalled bricks or stones can be made, although the treatment depends on the severity of the problem.

If spalling is localized, it is possible to cut out individual bricks or stones and replace them with matching ones. The sequence below describes how it's tackled with brickwork, but the process is similar for a stone wall.

Spalling bricks caused by frost damage

Priming brickwork

Brickwork will only need to be primed in certain circumstances. An alkali-resistant primer will guard against efflorescence (◁) and a stabilizing solution will bind crumbling masonry and help to seal it at the same time.

If you are planning to paint the wall for the first time with an exterior emulsion, you may find that the first coat is difficult to apply due to the suction of the dry, porous brick. Thin the first coat slightly with water.

To economize when using a reinforced emulsion (◁) prime the wall with a cement paint with a little fine sand mixed in thoroughly.

Waterproofing masonry

Colourless water-repellent fluids are intended to make masonry impervious to water without colouring it or stopping it from breathing (important to allow moisture within the walls to dry out).

Prepare the surface thoroughly before applying the fluid: make good any cracks in bricks or pointing and remove organic growth (◁) and allow the wall to dry out thoroughly.

Apply the fluid generously with a large paintbrush and stipple it into the joints. Apply a second coat as soon as the first has been absorbed to ensure that there are no bare patches where water could seep in. To be sure that you're covering the wall properly, use a sealant containing a fugitive dye, which disappears gradually after a few weeks.

Carefully paint up to surrounding woodwork; if you accidentally splash sealant onto it, wash it down immediately with a cloth dampened with white spirit.

If the area you need to treat is large, consider spraying on the fluid, using a hired spray gun. You'll need to rig up a substantial work platform (◁) and mask off all timber and metalwork that adjoins the wall. The fumes from the fluid can be dangerous if inhaled, so be sure to wear a proper respirator, which you can also hire.

Where the spalling is extensive, it's likely that the whole wall is porous and your best remedy is to paint on a stabilizing solution to bind the loose material together, then apply a textured wall finish, as used to patch pebbledash, which will disguise the faults and waterproof the wall at the same time.

Replacing a spalled brick
Use a cold chisel and club hammer to rake out the pointing surrounding the brick then chop out the brick itself. If the brick is difficult to prise out, drill into it many times with a large-diameter masonry bit, then attack the brick with a cold chisel and hammer: it should crumble, enabling you to remove the pieces easily.

To fit the replacement brick, first dampen the opening and spread mortar on the base and one side. Butter the dampened replacement brick on the top and one end and slot it into the hole (1).

Shape the pointing to match the surrounding brickwork then, once it is dry, apply a clear water repellent.

1 Replacing a spalled brick
Having mortared top and one end, slip the new brick into the hole you have cut.

Brickwork may be clad with a smooth or roughcast cement-based render for improved weatherproofing and to give a decorative finish; often the render is susceptible to the effects of damp and frost, which can cause cracking, bulging and staining. Before you redecorate a rendered wall, make good any damage, clean off surface dirt, mould growth and flaky material to achieve a long-lasting finish.

Cracked render allows moisture to penetrate

Blown pebbledash parts from the masonry

Rust staining due to leaky guttering

Repairing defective render

Before you repair cracked render, correct any structural faults which may have contributed to it. Brush to remove loose particles. Apply a stabilizing solution if the wall is dusty.

Ignore fine hair cracks if you paint the wall with reinforced emulsion, which covers minor faults. Rake out larger cracks using a cold chisel, dampen with water and fill flush with the surface with exterior filler. Fill major cracks with a mortar mix comprising I part cement: 4 parts builders' sand, with a little PVA bonding agent added to help it stick to the masonry.

Bulges in render can indicate that the cladding has parted from the masonry. Tap gently with a hammer to find the extent of these hollow areas; hack off the material to sound edges. Undercut the perimeter of the hole to give a grip for the filler material.

Brush out debris then apply a coat of PVA bonding agent. When PVA is tacky, trowel on a mortar mix as for filling cracks then smooth with a wet trowel.

Reinforcing a crack in render

To prevent a crack in render opening up again, reinforce the repair with a fine scrim embedded in a bitumen base coat. Rake out the crack to remove any loose material, then wet it. Fill the crack just proud of the surface with a mortar mix of I part cement: 4 parts builders' sand. When this has stiffened scrape it flush with the render.

When the mortar has hardened, brush on a generous coat of bitumen base coat, making sure it extends at least 75mm (3in) on both sides of the crack. Embed strips of open weave scrim (sold with the base coat) into the bitumen, using a stippling and brushing action (**1**). While it is still wet, feather the edges of the bitumen with a foam roller (**2**), bedding the scrim into it. After 24 hours, the bitumen will be hard, black and shiny. Apply a second coat, feather with a roller and, when it has dried, apply two full coats of a compatible reinforced emulsion (▷).

Patching pebbledash

Pebbledash comprises small stones stuck to a thin coat of render over a thicker base coat. If damp gets behind pebbledashing, one or both layers may separate. Hack off any loose render to a sound base and seal it with stabilizer. If necessary, repair the first 'scratchcoat' of render. Restore the texture of the top 'buttercoat' with a thick paste made from PVA bonding agent. Mix one part of cement paint powder with three parts clean, sharp (plastering) sand. Stir in one measure of bonding agent diluted with three parts water to form a thick, creamy paste. Load a banister brush and scrub the paste onto the bare surface.

Apply a second generous coat of paste, stippling it to form a coarse texture (**1**). Leave for about 15 minutes to firm up then, with a loaded brush, stipple it to match the texture of the pebbles. Let the paste harden fully before painting.

To leave the pebbledash unpainted, make a patch using replacement pebbles. The result will not be a perfect match but could save you painting the entire wall. Cut back the blown area and apply a scratchcoat followed by a buttercoat. While this is still wet, fling pebbles onto the surface from a dustpan; they should stick to the soft render but you'll have to repeat until the coverage is even.

Removing rust stains

Faulty plumbing will often leave rusty streaks on a rendered wall. Before decorating, prime the stains with an aluminium spirit-based sealer or they will bleed through. Rust marks may also appear on a pebbledashed wall, well away from any metalwork: these are caused by iron pyrites in the aggregate. Chip out the pyrites with a cold chisel, then seal the stain.

SEE ALSO

Details for: ▷	
Reinforced emulsion	41
Masonry paints	41
Rendering	36–39

I Embed the scrim

2 Feather with roller

I Stipple the texture

EXTERIOR RENDERING

Rendering is the application of a relatively thin layer of cement or cement-and-lime mortar to the surfaces of exterior walls to provide a decorative and weather-resistant finish.

Any such treatment of exterior walls should be carefully considered beforehand, because the finished outer surface should always harmonize with the character of a building and not look ill at ease with those of its neighbours. This is particularly important in the case of terraced housing, where the fronts of the houses form an unbroken run of wall.

Planning ahead

There are no regulations controlling the change of colour or texture of outer walls except those of listed buildings and consequently one often sees houses made conspicuous by their individualistic decorative treatment.

Rerendering a wall would always be acceptable as it is merely a case of renewing what is already there.

Rendering old brickwork might improve its weather-resistance, but at the cost of destroying the appearance of the building. Here it would be better to rake out the mortar joints, repoint them and, if necessary, treat the brickwork with a clear sealant (◁).

Rendering techniques

The technique used in rendering is virtually the same as in plastering, for which cement and lime are also sometimes used, and it generally involves using the same tools, though a wooden float is better than a plasterer's trowel for finishing the cement rendering. The wood leaves a finely textured surface that looks better than the very smooth one produced with a metal trowel.

Rendering the walls of a house is really a job for the professional, as it involves covering a large area and colour-matching the batches of mortar, which is critical if the finished job is not to look patchy.

While a non-professional can undertake repairs to rendering it is very difficult to match the colour of the new work to the old. You might consider finishing the complete wall with paint.

New work can be approached by dividing the wall into manageable panels, using screed battens as in plastering (◁). But colour matching the mortar will still be a problem, and 'losing' the joints can be difficult. It might pay to concentrate on getting the rendering flat and then disguising any patchiness with paint – here again arises the question of the character of the house.

Before attempting to render a large wall, practise if possible on a smaller project, such as a garden wall.

BINDERS FOR EXTERIOR RENDERING

MORTAR

Mortar is a mixture of sand, cement and clean water. The sand gives the mix bulk and the cement binds the particles. A cement mix will bond to any ordinary masonry material and to metal. Mortars of various strengths are produced by adjusting the proportions of sand and cement, or by adding lime.

A mortar must not be stronger than the materials onto which it is being applied. In a wall of dense, hard bricks a cement-and-sand mix can be used. For soft bricks or blocks a weaker mix of cement, sand and lime is appropriate.

Cement sets by a chemical reaction with water known as hydration, and begins as soon as the water is added. Cement does not need to dry out in order to set, and the more slowly it dries out the stronger it will be.

Normally an average mix will stay workable for at least two hours. It will continue to gain strength for a few days after its initial set, reaching full strength in about a month.

Hot weather will reduce the workable time and can affect the set of the mortar by making it dry out too fast. In these conditions the work should be kept damp by being lightly sprayed with water or by being covered with polythene sheeting to retain the moisture and slow the drying time.

CEMENT

Cement made from limestone or chalk and clay is generally called Portland cement. Various types are made by adding other materials or by modifying the production methods.

Ordinary Portland cement (OPC)

This common light grey cement is mixed with aggregates for concrete and mortars. It is available in 50kg (110lb) bags but can be had in smaller amounts.

White Portland cement

This is similar to ordinary Portland but is white and more expensive. It makes light-coloured mixes for bricklaying, rendering and concrete.

Coloured cement

Some coloured cements are available and also pigments for colouring mortar mixes. The materials must be carefully proportioned for the batches to match.

Quick-setting cement

This cement is mixed with water and sets hard in 30 minutes. It is non-shrinking and waterproof; useful for small repair jobs.

Masonry cement

Specially made for rendering and bricklaying, this grey cement is not suitable for concrete. Add no lime to it.

LIME

Lime is made from limestone or chalk. When it leaves the kiln it is called quicklime, and may be non-hydraulic or hydraulic. Non-hydraulic lime, in use generally, sets by combining with carbon dioxide from the air as the water mixed with it dries out. Hydraulic lime has similar properties to cement; it sets when water is added and so can be used under water. When quicklime is 'slaked' – mixed with water – it expands and gives off heat. This is particularly strong with non-hydraulic quicklime.

Lime must be properly slaked before use, and at one time a batch would be soaked in a tub weeks beforehand. This soaked lime was called lime putty.

Pre-slaked non-hydraulic lime powder, or hydrated lime, is sold by builders' merchants. It can be used at once, but is often soaked 24 hours before use to make lime putty. The lime is mixed with water to a creamy consistency or with sand and water and left to stand as lime mortar called 'coarse stuff'. This can be kept for some days without setting if it is heaped up and covered with polythene sheet to prevent the water evaporating.

The less active hydraulic lime is dry-mixed with the sand like cement powder and the slaking process takes place when the water is added.

AGGREGATES

Mortars are mixed with the finest aggregate, sand, and sand is graded by the size and shape of its particles. A well-graded sand will have particles of different sizes, not ones which are uniformly large or small.

Types of sand

'Sharp' sand is used with other, coarse aggregates for making concrete and floor screeds. Plasterers' sharp sand is of a finer grade, and is used for rendering. 'Soft', 'builders' or 'bricklayers' sand has smoother particles and is used for masonry work.

Use well-washed sand, as impurities can weaken a mortar and affect the set. A good sand should not stain your hand if you squeeze it.

Most aggregates may be bought from builders' merchants by the cubic metre; some suppliers sell it in small packs.

Stone chippings

Specially prepared crushed stone in various colours is used for pebble-dash rendering. Buy enough for the whole job in hand (check this with your supplier) as additional stones from another batch may not match the colour of those you have.

If you do start running short, stop work at a corner rather than part way across a wall. The extra stones can be mixed with the remaining ones and a subtle change of colour is less likely to show on the adjacent wall.

Dry-mixed mortar

Pre-packed sand-and-cement mortar mixes are sold by builders' merchants and DIY shops in large and small packs. They are ready-proportioned for different kinds of application and require only water to be added. As sand and cement 'settle out', a whole bag should be used and mixed well, before adding the water.

This way of buying sand and cement is more expensive than buying loose material but the convenience and low wastage are great advantages, especially for small jobs.

STORING SAND AND CEMENT

Storing the materials should not normally be necessary because it is best that they should be bought as required and used up by the end of the job.

However, if you are held up for a time after taking delivery, you should store powder or pre-mixed materials on plastic sheet in a dry space. Sand should be stored in a neat heap on a board or a plastic sheet and protected from windblown dirt and rain with plastic sheeting.

Storing sand ▶
Dirty sand can affect the set of the cement. Keep it covered with plastic sheeting.

ADDITIVES

Proprietary additives which modify the properties of mortar are added to the mix in precise proportions according to manufacturers' instructions. Their functions vary. Waterproofers, which make mortar impervious by sealing its pores, may be used when rendering on exposed walls. Plasticisers, additives which make a mortar easier to work, can be used instead of lime.

SEE ALSO

Details for: ▷

| Mixing mortar | 38 |

MORTAR MIXES FOR RENDERING

The mix for a mortar will depend on the strength of the material being rendered and the degree of its exposure. The undercoat mix should not be stronger than the background, and the top coat should be no stronger than the undercoat.

Though these considerations are not critical in most DIY work with mortar, where a situation does dictate that a precise mix is required, the proportions of the materials must be measured quite accurately.

The material is measured by volume, using a bucket. Loose cement and damp sand tend to 'bulk up' when loaded. Cement powder can be made to settle by tapping the bucket, and it is then topped up. Damp sand will not settle, so the measure is usually increased by 25 per cent. Dry sand and saturated sand will settle to a normal measure.

| TYPE OF BACKGROUND | TYPE OF MIX | PARTS BY VOLUME | | | |
| | | AVERAGE EXPOSURE | | SEVERE EXPOSURE | |
		UNDERCOAT	TOP COAT	UNDERCOAT	TOP COAT
Low suction backgrounds: Hard dense clay bricks Dense concrete blocks Stone masonry Normal ballast concrete	Cement: lime: sand	1-½-4½	1-1-6*	1-½-4½	1-1-6
	Cement: sand & plasticiser	1-4	1-6*	1-4	1-6
	Masonry cement: sand	1-3½	1-5*	1-3½	1-5
Normal absorption backgrounds: Most average strength bricks Clay blocks Normal concrete blocks Aerated concrete blocks	Cement: lime: sand	1-1-6	1-2-9*	1-1-6	1-1-6
	Cement: sand & plasticiser	1-6	1-8*	1-6	1-6
	Masonry cement: sand	1-5	1-6½*	1-5	1-5

*Suitable mixes for interior plastering undercoats

MIXING MORTAR FOR RENDERING

1 Shovel dry mix from one heap to another

2 Form a well in the heap and pour in water

3 Shovel dry material into water and mix

4 Sprinkle mix with water if too dry

Mix only as much mortar as you can use in an hour, and if the weather is very hot and dry shorten this to half an hour. Keep all your mixing tools and equipment thoroughly washed so that no mortar sets on them.

Measure the required level bucketfuls of sand onto the mortar board (◁) or – for larger quantities – onto a smooth, level base such as a concrete drive.

Using a second dry bucket and shovel, kept exclusively for cement powder, measure out the cement, tapping the bucket to settle the loose powder and topping it up as needed. Tip the cement over the heaped-up sand and mix sand and cement together by shovelling them from one heap to another and back again (1), and continue to turn this dry mix – the sand will actually be damp – until it takes on a uniform grey colour.

Form a well in the centre of the heap and pour in some water (2) – but not too much at this stage.

Shovel the dry mix from the sides of the heap into the water until the water is absorbed (3). If you are left with dry material add more water as you go until you achieve the right firm, plastic consistency in the mortar, turning it repeatedly to mix it thoroughly to an even colour. It is quite likely that you will misjudge the amount of water at first so if after turning the mix it is still relatively dry sprinkle it with water (4). But remember that too much water will weaken the mix.

Draw the back of your shovel across the mortar with a sawing action to test its consistency (5). The ribs formed in the mixture by this action should not slump back or crumble, which would indicate respectively that it is either too wet or too dry. The back of the shovel should leave a smooth texture on the surface of the mortar.

Make a note of the amount of water used in proportion to the dry materials so that further mixes will be consistent.

For cement-lime-sand mixes the lime powder can be added with the cement and dry-mixed as described above. Otherwise lime putty can be mixed with the sand before the cement is added, or the cement can be added to prepared 'coarse stuff' (◁). When you have finished, hose down and sweep clean the work area particularly if it is a driveway as any remaining cement slurry will stain the surface.

You can hire a small-capacity electric or petrol-driven cement mixer. Such a mixer can save you much time and effort, especially on big jobs, and is quite easy to use.

Set the machine as close as possible to the work area and place a board under the drum to catch any spilt materials. If it is an electric-powered machine take all due precautions with the power supply and keep the cables well clear of the work.

Load the drum with half the measure of sand and add a similar proportion of cement, and lime if required. Dry-mix them by running the mixer, then add some water.

Load the remainder of the materials in the same sequence, adding a little water in between.

Run the mixer for a couple of minutes to mix the materials thoroughly, then stop the machine and test some of the mix for consistency, as it may appear stiffer than it is.

A rendering mix should be less workable than a mix for bricklaying, and for rendering blockwork it is usually stiffer than for brickwork, but this will depend on the absorbency of the wall.

It is advisable to wash out the drum of the mixer after each mix and to scour it out with water and some coarse aggregate at the end of the working day.

If you return the machine with dry or drying mortar in its drum you may be charged extra.

Cement mixer
Hire an electric or petrol-driven mixer when a large batch of mortar is required

5 Test consistency of mix with shovel

TEXTURED RENDERINGS

Renderings may be textured by tooling while damp to make patterns or by applying a coarse aggregate, which is a fairly skilled procedure (See below for details). Reproduce a texture when patch repairing.

Roughcast rendering
For this rendering, mix aggregate no more than 10mm (⅜in) in size with the top coat mortar. Use about half as much as the sand used with enough water for a sticky mix. Flick it on the wall to build up an even coat.

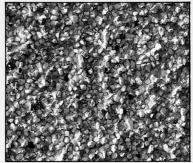

Pebbledash rendering
Crushed stone aggregate gives pebbledash its colour, and an even distribution of the chippings is required to avoid patchiness.
A 10 to 12mm (⅜ to ½in) top coat is applied and the stones thrown at it while soft, then pressed with a float to bed them in.

Tyrolean finish
A fine cement mix is sprayed from a hand-cranked 'Tyrolean machine' to build up a decorative honeycomb texture over a dry undercoat rendering. Doors, windows, gutters etc., must be masked beforehand. Tyrolean machines may be hired.

APPLYING RENDERINGS

Preparing the surface

Neatly chop away the old loose coating on areas of cracked or blistered rendering, using a hammer and chisel. Rake out the mortar joints in the exposed brickwork if necessary and brush the area down. Clean off any organic growth like lichen or algae and apply a fungicide (▷).

Working platform

Set up a safe working platform from which to do rendering. You will need both hands free to use the tools, so it cannot be done from a ladder. Pairs of steps with a scaffold board between them can be used for working on ground floor walls, but for upper walls you will need a scaffold tower (▷).

SEE ALSO

Details for: ▷	
Work platform	78
Organic growth	32
Render	35

New work

Set up 10mm (⅜in) vertical screed battens spaced no more than 900mm (3ft) apart, fixing them with masonry nails into the mortar joints of the brickwork. Check them for level and pack them out where necessary.

Between two battens apply undercoat rendering with a firm pressure to make it bond well onto the background, building it up to the thickness of the screed battens.

Level the mortar off with a straight-edge laid across the battens and worked upward with a side-to-side movement, then scratch the surface of the mortar to provide a key for the top coat and leave it to set for a week.

You can fill in the panels between the battens in sequence or fill alternate ones. Whichever you do, let them set before removing the screeds. The former method will leave narrow strips to be filled in where the screeds have been removed, whereas in the second method the newly set alternate panels of rendering will provide a levelling surface for those to be done.

Apply top coat rendering about 6mm (¼in) thick, either freehand or with the aid of screed battens as before. In any case use a straightedge for levelling it off, then finish it with a wooden float, which will produce a finely textured surface on the rendering.

**Applying rendering
Order of working**
1 Set up a safe work platform
2 Divide the wall with vertical screeds
3 Apply the undercoat rendering between the screeds or alternate panels
4 Remove screeds and fill in gaps or panels
5 Apply top coat over keyed undercoat

PATCH REPAIRS

The rendering should be applied with a metal plasterer's trowel and the top coat finished with a wooden float.

Take a trowelful of mortar from your hawk and spread it on the wall with a firm pressure and applying it with an upward stroke (**1**).

Build up the undercoat layer no more than two thirds the thickness of the original rendering or 10mm (⅜in), whichever is the thinner.

Level the mortar with a straight-edged batten that fits within the cut-out of the area being patched, then scratch a key in the surface for the top coat (**2**). Leave the undercoat to set and strengthen for a few days.

The top coat
Before applying the top coat dampen the undercoat rendering if necessary to even out the suction. When the coat is on, level it with a straightedge laid across the surfaces of the surrounding rendering and worked upward with a side-to-side motion.

1 Use firm pressure

2 Key the surface

SAFETY WHEN PAINTING

Decorating isn't dangerous so long as you take sensible precautions to protect your health.

- **Ensure good ventilation indoors while applying a finish and when it is drying. Wear a facemask if you have respiratory problems.**
- **Do not smoke while painting or in the vicinity of drying paint.**
- **Contain paint spillages outside with sand or earth and don't allow it to enter a drain.**
- **If you splash paint in your eyes, flush them with copious amounts of water with your lids held open; if symptoms persist, visit a doctor.**
- **Wear barrier cream or gloves on sensitive hands. Use a proprietary skin cleanser to remove paint from the skin or wash it off with warm soapy water. Do not use paint thinners to clean your skin.**
- **Keep any finish and thinners out of reach of children. If a child swallows a substance, do not attempt to make it vomit but seek medical treatment.**

Strain old paint
If you're using leftover paint, filter it through a piece of muslin or old tights stretched over the rim of a container.

Resealing the lid
Wipe the rim of the can clean before you replace the lid, then tap it down all round with a hammer over a softwood block.

PREPARING THE PAINT

Whether you're using newly purchased paint or leftovers from previous jobs, there are some basic rules to observe before you apply it.
- Wipe dust from the paint can, then prise off the lid with the side of a knife blade. Don't use a screwdriver: it only buckles the edge of the lid, preventing an airtight seal and making subsequent removal difficult.
- Gently stir liquid paints with a wooden stick to blend the pigment and medium. There's no need to stir thixotropic paints unless the medium has separated; if you have to stir it, leave it to gel again before using.
- If a skin has formed on paint, cut round the edge with a knife and lift out in one piece with a stick. It's a good idea to store the can on its lid, so that a skin cannot form on top of the paint.
- Whether the paint is old or new, transfer a small amount into a paint kettle or plastic bucket. Old paint should be filtered at the same time, tying a piece of muslin or old nylon tights across the rim of the kettle.

PAINTING EXTERIOR MASONRY

The outside walls of your house need painting for two major reasons: to give a clean, bright appearance and to protect the surface from the rigours of the climate. What you use as a finish and how you apply it depends on what the walls are made of, their condition and the degree of protection they need. Bricks are traditionally left bare, but may require a coat of paint if they're in bad condition or previous attempts to decorate have resulted in a poor finish. Rendered walls are normally painted to brighten the naturally dull grey colour of the cement; pebbledashed surfaces may need a colourful coat to disguise previous conspicuous patches. On the other hand, you may just want to change the present colour of your walls for a fresh appearance.

Working to a plan

Before you start painting the outside walls of your house, plan your time carefully. Depending on the preparation even a small house will take a few weeks to complete.

It's not necessary to tackle the whole job at once, although it is preferable – the weather may change to the detriment of your timetable. You can split the work into separate stages with days (even weeks) in between, so long as you divide the walls into manageable sections. Use window and door frames, bays, downpipes and corners of walls to form break lines that will disguise joins.

Start at the top of the house, working right to left if you are right-handed (vice versa if you are left-handed).

● Black dot denotes compatibility. All surfaces must be clean, sound, dry and free from organic growth.

FINISHES FOR MASONRY

	Cement paint	Exterior emulsion paint	Reinforced emulsion paint	Spirit-thinned masonry paint	Textured coating	Floor paint
SUITABLE TO COVER						
Brick	●	●	●	●	●	●
Stone	●	●	●	●	●	●
Concrete	●	●	●	●	●	●
Cement rendering	●	●	●	●	●	●
Pebbledash	●	●	●	●	●	
Asbestos cement	●		●	●	●	●
Emulsion paint		●	●	●	●	
Oil-based paint				●	●	●
Cement paint	●	●	●	●	●	●
DRYING TIME: HOURS						
Touch dry	1-2	1-2	2-3	1-2	6	2-3
Re-coatable	24	4	24	24	24-48	12-24
THINNERS: SOLVENTS						
Water-thinned	●	●	●		●	
White spirit-thinned				●		●
NUMBER OF COATS						
Normal conditions	2	2	1-2	2	1	1-2
COVERAGE: DEPENDING ON WALL TEXTURE						
Sq.metres per litre		4-10	3-6.5	3-6		5-15
Sq.metres per kg	1.5-3.5				1-2	
METHOD OF APPLICATION						
Brush	●	●	●	●	●	●
Roller	●	●	●	●	●	
Spray gun	●	●	●	●		●

SUITABLE PAINTS FOR EXTERIOR MASONRY

There are various grades of paint suitable for decorating and protecting exterior masonry, which take into account economy, standard of finish, durability and coverage. Use the chart opposite for quick reference.

CEMENT PAINT

Cement paint is supplied as a dry powder, to which water is added. It is based on white cement but pigments are added to produce a range of colours. Cement paint is the cheapest of the paints suitable for exterior use, although it is not as weatherproof as some others. Spray new or porous surfaces with water before you apply two coats.

Mixing cement paint
Shake or roll the container to loosen the powder, then add two volumes of powder to one of water in a clean bucket. Stir it to a smooth paste then add a little more water until you achieve a full-bodied, creamy consistency. Mix up no more than you can use in one hour, or it will start to dry.

Adding an aggregate
When you're painting a dense wall or one treated with a stabilizing solution so that its porosity is substantially reduced, it is advisable to add clean sand to the mix. It also provides added protection for an exposed wall and helps to cover dark colours. If the sand changes the colour of the paint, add it to the first coat only. Use one volume of sand to four of powder, but stir it in when the paint is still in its paste-like consistency.

EXTERIOR-GRADE EMULSION

Exterior-grade emulsion resembles the interior type; it is water-thinnable and dries to a similar smooth, matt finish. However, it is formulated to make it weatherproof and includes an additive to prevent mould growth; so apart from reinforced emulsions, it is the only emulsion paint recommended for use on outside walls.

The paint is ready for use but thin the first coat on porous walls with 20 per cent water. Follow up with one or two full-strength coats (depending on the colour of the paint).

REINFORCED EMULSION

Reinforced emulsion is a water-thinnable, resin-based paint to which has been added powdered mica or a similar fine aggregate. It dries with a textured finish that is extremely weatherproof, even in coastal districts or industrial areas where darker colours are especially suitable.

Although cracks and holes must be filled prior to painting, reinforced emulsion will cover hair cracks and crazing. Apply two coats of paint in normal conditions but you can economize by using sanded cement paint for the first coat.

SPIRIT-THINNED MASONRY PAINT

A few masonry paints suitable for exterior walls are thinned with white spirit but they are based on special resins so that, unlike most oil-based paints, they can be used on new walls without priming first with an alkali-resistant primer (▷). Check with manufacturer's recommendations. However, it is advisable to thin the first coat with 15 per cent white spirit.

TEXTURED COATING

A thick textured coating can be applied to exterior walls. It is a thoroughly weatherproof, self-coloured coating, but it can be overpainted to match other colours. The usual preparation is necessary and brickwork should be pointed flush. Large cracks should be filled, but a textured coating will cover fine cracks. The paste is brushed or rolled onto the wall, then left to harden, forming an even texture. On the other hand, you can produce a texture of your choice using a variety of simple tools. It's an easy process, but practise on a small section first.

Concrete floor paints

Floor paints are specially prepared to withstand hard wear. They are especially suitable for concrete garage or workshop floors, but they are also used for stone paving, steps and other concrete structures. They can be used inside for playroom floors.

The floor must be clean and dry and free from oil or grease. If the concrete is freshly laid, allow it to mature for at least three months before painting. Thin the first coat of paint with 10 per cent white spirit.

Don't use floor paint over a surface sealed with a proprietary concrete sealer, but you can cover other paints so long as they are keyed first.

The best way to paint a large area is to use a paintbrush around the edges, then fit an extension to a paint roller for the bulk of the floor.

SEE ALSO

Details for: ▷
Alkali-resistant
primer   78
Preparing masonry   32–35

Apply paint with a roller on an extension

Paint in manageable sections
You can't hope to paint an entire house in one session, so divide each elevation into manageable sections to disguise the joins. The horizontal moulding divides the wall neatly into two sections, and the raised door and window surrounds are convenient break lines.

TECHNIQUES FOR PAINTING MASONRY

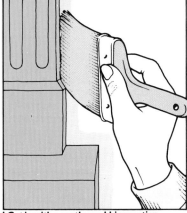

1 Cut in with a gentle scrubbing motion

3 Use a banister brush
Tackle deeply textured wall surfaces with a banister brush, using a scrubbing action.

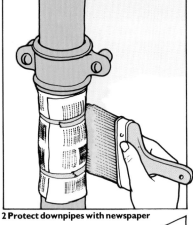

2 Protect downpipes with newspaper

4 Use a roller
For speed in application, use a paint roller with a deep pile for heavy textures, a medium pile for light textures and smooth wall surfaces.

5 Spray onto the apex of external corners

6 Spray internal corners as separate surfaces

Using paintbrushes

Choose a 100 to 150mm (4 to 6in) wide paintbrush for walls; larger ones are heavy and tiring to use. A good-quality brush with coarse bristles will last longer on rough walls. For a good coverage, apply the paint with vertical strokes, criss-crossed with horizontal ones. You will find it necessary to stipple paint into textured surfaces.

Cutting in

Painting up to a feature such as a door or window frame is known as cutting in. On a smooth surface, you should be able to paint a reasonably straight edge following the line of the feature, but it's difficult to apply the paint to a heavily textured wall with a normal brush stroke. Don't just apply more paint to overcome the problem; instead, touch the tip of the brush only to the wall, using a gentle scrubbing action (1), then brush excess paint away from the feature once the texture is filled.

Wipe splashed paint from window and door frames with a cloth dampened with the appropriate thinner.

Painting behind pipes

To protect rainwater downpipes, tape a roll of newspaper around them. Stipple behind the pipe with a brush then slide the paper tube down the pipe to mask the next section (2).

Painting with a banister brush

Use a banister brush (3) to paint deep textures such as pebbledash. Pour some paint into a roller tray and dab the brush in to load it. Scrub the paint onto the wall using circular strokes to work it well into the uneven surface.

Using a paint roller

A roller (4) will apply paint three times faster than a brush. Use a deep-pile roller for heavy textures or a medium-pile for lightly textured or smooth walls. Rollers wear quickly on rough walls, so have a spare sleeve handy. Vary the angle of the stroke when using a roller to ensure an even coverage and use a brush to cut into angles and obstructions.

A paint tray is difficult to use at the top of a ladder, unless you fit a tool support, or better still erect a flat platform to work from (◁).

Using a spray gun

Spraying is the quickest and most efficient way to apply paint to a large expanse of wall. But you will have to mask all the parts you do not want to paint, using newspaper and masking tape. The paint must be thinned by about 10 per cent for spraying: set the spray gun according to the manufacturer's instructions to suit the particular paint. It is advisable to wear a respirator when spraying.

Hold the gun about 225mm (9in) away from the wall and keep it moving with even, parallel passes. Slightly overlap each pass and try to keep the gun pointing directly at the surface – tricky while standing on a ladder. Trigger the gun just before each pass and release it at the end of the stroke.

When spraying a large, blank wall, paint it into vertical bands overlapping each band by 100mm (4in).

Spray external corners by aiming the gun directly at the apex so that paint falls evenly on both surfaces (5). When two walls meet at an internal angle, treat each surface separately (6).

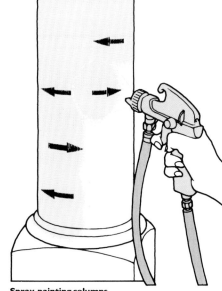

Spray-painting columns
Columns, part of a front door portico, for instance, should be painted in a series of overlapping vertical bands. Apply the bands by running the spray gun from side to side as you work down the column.

Most pitched roofs were once built on site from individual lengths of timber, but nowadays, for economy of time and materials, many builders use prefabricated frames called trussed rafters. These are specifically designed to meet the loading needs of a given house and, unlike traditional roofs, are usually not suitable for conversion because to remove any part of the structure can cause it to collapse.

Close couple roof
A roof structure which has its rafters joined by a tie member. A variation is the collar roof where the ties (collars) are set at a higher level.

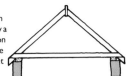

Basic construction

The framework of an ordinary pitched roof is based on a triangle, the most rigid and economical form for a load-bearing structure. The weight of the roof covering is carried by the sloping members, the 'common rafters', which are set in opposing pairs whose heads meet against a central 'ridge board'. The lower ends, or feet, of the rafters are fixed to timber wall plates which are bedded on the exterior walls and distribute the weight uniformly.

To stop the roof's weight pushing the walls out, horizontal 'ties' are fixed to the ends of each pair of rafters and the wall plates, forming a simple structure called a 'close couple' roof. The ties (ceiling joists) also usually support the ceiling plaster. The rafters are also linked by the roofing battens.

PITCHED ROOF TYPES

SINGLE ROOFS

Any roof, pitched or flat, with unbraced rafters – except the pitched roof's ties – is called a single roof (**1**), and is only suitable for light coverings and short spans. For a wide span or heavy covering the design would need unduly large roof timbers.

DOUBLE ROOFS

In double roofs horizontal beams called 'purlins' support the rafters (**2**), linking them either midway between foot and ridge or 2.5m (8ft) apart. This reduces the span of the rafters and allows lighter timber to be used, 100 × 50mm (4 × 2in) being common. The purlins' section depends on the weight of the roof covering, but it usually exceeds that of the rafters; 200 × 50mm (8 × 2in) is normal. The purlins' ends are supported on the brickwork of a gable wall or, in a hipped roof, by the hip rafters.

To keep the purlins' size down, struts may be set in opposing pairs to brace them diagonally at every fourth or fifth pair of rafters. Struts transfer some weight back to the centre of the ceiling joists, which are supported there by a load-bearing dividing wall at right angles to them.

The ends of the struts may be jointed over a horizontal 'binder' fixed to the joists right above the supporting wall.

Where ceiling joists are fairly light and the span could make them sag, vertical timber 'hangers' are fixed at the top to every third or fourth rafter or to adjacent purlins at like intervals, and at the bottom to a 'binder' fixed at right angles across the joists.

TRUSSED ROOFS

Some traditional roofs embody trusses, rigid triangular frames that replace load-bearing partition walls to give a wider span. Trusses carry the purlins, which in turn support the rafters and form a 'triple' or 'framed' roof. The trusses are spaced at 1.8m (6ft) or more depending on the purlins' section or the roof covering's weight. As main bearers for the roof they transmit weight to the exterior walls. Few trussed roofs can be converted and you should not try to cut into them.

'Trussed rafters' are now used in all new housing (**3**). Computer-designed for economy plus rigidity, the trusses are prefabricated of planed softwood 38mm (1½in) thick and up to 150mm (6in) wide, depending on roof loading. Each truss combines common rafters, tie and strut bracing in one frame; the members are butt jointed and fixed with special nailed plate connectors.

The trusses are spaced a maximum 600mm (2ft) apart, linked horizontally with bracing members nailed to the struts. Such roofs are quite light, and usually fixed to the walls with steel anchor straps to resist wind pressure.

1 Single roof
The basic construction for a pitched roof
1 Common rafter
2 Tie
3 Wall plate
4 Ridge board

2 Double roof
The most common traditional roof construction sometimes referred to as a purlin roof.
1 Common rafter
2 Tie
3 Wall plate
4 Ridge board
5 Purlin
6 Strut
7 Binder
8 Hanger

3 Trussed-rafter roof
Each prefabricated truss combines the common rafters, tie and struts in one frame. Diagonal and horizontal bracing is used to join them together but a ridge board is not fitted.
1 Trussed rafter
2 Wall plate
3 Bracing

ROOF ELEMENTS

Hips and valleys

All the components mentioned earlier (◁) are found in ordinary gable roofs. One with a hipped end or valleys has other parts, but all follow the same principles. Here a double roof shows the parts and their names.

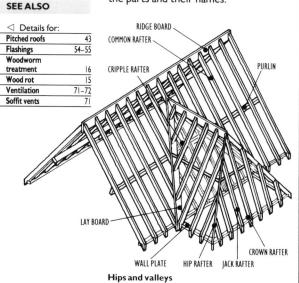

RIDGE BOARD
COMMON RAFTER
CRIPPLE RAFTER
PURLIN
LAY BOARD
CROWN RAFTER
WALL PLATE   HIP RAFTER   JACK RAFTER

Hips and valleys

Eaves

The overhang of rafters past the outer walls is called the eaves, but sometimes rafters are cut flush with the walls and a fascia board along their ends protects them and supports the guttering **(1)**. Projecting rafters can be left open, the ends exposed **(2)**, and gutter brackets screwed to their sides or top edge.

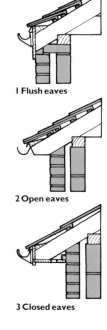

1 Flush eaves

2 Open eaves

3 Closed eaves

Closed eaves

The back of a fascia is usually grooved to take a soffit board, which closes the eaves **(3)**. The board can be at 90 degrees to the wall or slope with the rafters, and it can be of various weatherproof materials. If loft insulation is laid, a roof with closed eaves must be ventilated by a small gap between soffit board and wall or by fitted vents (◁).

The verge

The verge is the sloping edge of the roofing and can end flush with the wall or project past it. A flush verge means a roof structure that stops at the wall, end rafters placed close to it and the roofing overlying it **(4)**. A projecting verge means timbers extending beyond the wall and short lengths of rafter set in the brick to carry an outer rafter with a 'barge board' fixed to it. Behind the barge board may be a soffit board to conceal the outer rafters **(5)**.

ROOF STRUCTURE PROBLEMS

A roof structure can fail when timbers decay through inadequate weatherproofing, condensation or insect attack. It can also result from overloading caused by too-light original timbers, a new roof covering of heavier material or the cutting of a window opening that is not properly braced. You can detect any movement of a roof structure from outside. From ground level any sagging of the roof will be seen in the lines of the roof covering.

INSPECTION

The roof should also be inspected from inside. In any case, this should be done annually to check the weathering and for freedom from woodworm infestation.

Work in a good light. If your loft has no lighting use a mains-powered lead light. In an unboarded attic place strong boards across the joists to walk on.

ROT AND INFESTATION

Rot in roof timbers is a serious problem which should be put right by experts, but its cause should also be identified and promptly dealt with.

Rot is caused by damp conditions that encourage wood-rotting fungi (◁) to grow. Inspect the roof covering closely for loose or damaged elements in the general area of the rot, though on a pitched roof water may be penetrating the covering at a higher level and so not be immediately obvious. If the rot is close to a gable wall you should suspect the flashing (◁). Rot can also be caused by condensation, the remedy for which is usually better ventilation (◁).

If you bring in contractors to treat the rot it is better to have them make all the repairs. Their work is covered by a guarantee which may be invalidated if you attempt to deal with the cause yourself to save money.

Wood-boring beetle infestation should also be dealt with by professionals if it is serious. Severely infected wood may have to be cut out and replaced, and anyway the whole structure will have to be spray treated (◁).

STRENGTHENING THE ROOF

A roof that shows signs of sagging may have to be braced, though it may not be necessary if a sound structure has stabilised and the roof is weatherproof. In some old buildings a slightly sagging roof line is considered attractive.

Consult a surveyor if you suspect a roof is weak. Apart from a sagging roof line, the walls under the eaves should be inspected for bulging and checked with a plumb line. Bulging may occur where window openings are close to the eaves, making a wall relatively weak. It may be due to the roof spreading because of inadequate fixing of the ties and rafters. If this is so, call in a builder or roofing contractor to do the repair work.

A lightly structured roof can be made stronger by adding extra timbers. The method chosen will depend on the type of roof, its span, its loading and its condition. Where the lengths or section of new timbers are not too large the repair may be possible from inside. If not, at least some of the roof covering may have to be stripped off. Any given roof must be surveyed and the most economical solution adopted.

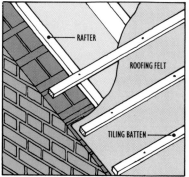

RAFTER
ROOFING FELT
TILING BATTEN

4 Flush verge

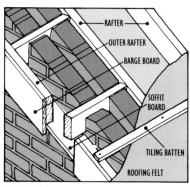

RAFTER
OUTER RAFTER
BARGE BOARD
SOFFIT BOARD
TILING BATTEN
ROOFING FELT

5 Projecting verge

TYPES OF ROOFING

Roof coverings are manufactured by moulding clay or concrete into various profiles or by cutting natural materials such as slate into flat sheets.

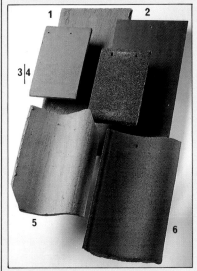

Selection of roof covering materials
1 Natural slate. 2 Machine-made slate. 3 Plain tile (clay). 4 Plain tile (concrete). 5 Plain pantile (clay). 6 Interlocking pantile (concrete).

PITCHED ROOF COVERINGS

Coverings for domestic pitched roofs follow a long tradition, and despite the developments in new materials the older ones and the ways of using them have not changed radically.

Like most early building materials, those used for roofing were generally of local origin, which led to a diversity of roof coverings, including tiles, slates and timber shingles. For centuries they were hand-made, and had their own characteristics, to be seen in various regional styles.

During the last century the more durable roof coverings, such as tiles and slates, became more widely adopted.

Most roofing materials are laid across the roof in rows called courses so that the bottom edge of each overlaps the top of the one below. This means that they are laid working from the eaves up the slope of the roof to the ridge.

Specially shaped tiles are used for capping the ridge or hips so as to weatherproof the junctions of the slopes. Where the covering meets a chimney or a wall it is protected with flashing, usually made of lead or mortar.

Roof-covering components for a pitched roof.
1 Tile or slate covering
2 Ridge tile
3 Gable end
4 Projecting verge
5 Barge board
6 Eaves
7 Fascia
8 Soffit
9 Hipped end
10 Hip tile
11 Valley
12 Flush verge
13 Lead stepped flashing
14 Back gutter
15 Apron

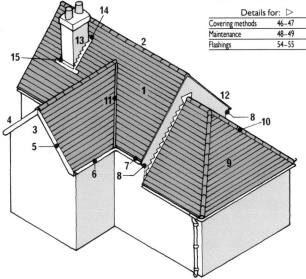

SEE ALSO

Details for: ▷	
Covering methods	46–47
Maintenance	48–49
Flashings	54–55

TYPICAL COVERINGS FOR PITCHED ROOFS

Material	Common sizes	Finish	Colour	Fixing	Weight* Kg/M² (1lb/sq yd)	Minimum pitch Degrees
TYPE OF COVERING						
SLATE Split metamorphic sedimentary rock	Lengths: 300 to 600 mm (1 ft to 2 ft) Widths: 180 to 350 mm (7 in to 1 ft 2 in)	Natural	Natural Blue Grey Green	Two nails	27.5 to 70 (50.69 to 129)	17½°
MACHINE-MADE SLATE Non-asbestos cement Asbestos cement	Lengths: 400, 500, 600 mm (1 ft 4 in, 1 ft 8 in, 2 ft) Widths: 200, 250, 300, 350 mm (8in, 10in, 1ft, 1ft 2in)	Acrylic coating	Grey Blue Brown	Two nails plus copper disc rivet in tail	18.5 to 20.8 (34.10 to 38.34)	17½°
STONE Split sandstone or limestone sedimentary rock	Random and as natural slate	Natural	Natural Yellow Grey Green	Two nails	90 (165.9)	17½°
SHINGLES Split or sawn red cedar	Length: 400 mm (1 ft 4 in) Widths: 75 to 300 mm (3 in to 1 ft)	Natural	Natural Brown Grey	Two nails	7 (12.83)	20°
PLAIN TILES Hand- or machine-moulded clay or machine-made concrete	Length: 265 mm (10½ in) Width: 165 mm (6½ in)	Sanded Smooth	Red Brown Grey Blue Green	Two nails or loose laid on nibs	65 (119.8)	40° clay 35° concrete
INTERLOCKING TILES Hand- or machine-moulded clay or machine-made concrete	Lengths: 380, 410, 430 mm (1ft 3in, 1ft 4½in, 1ft 5in) Widths: 220, 330, 380 mm (9in, 1ft 1in, 1ft 3in)	Sanded Smooth Glazed	Red Brown Grey Blue Green	Loose laid on nibs, nailed or clipped	40 to 57 (73.73 to 105.07)	35° clay 17½° to 30° concrete

*Approx

TYPES OF ROOF COVERING

All roof coverings built up with small overlapping units fall broadly into two classifications: 'double-lap' and 'single-lap', categories indicating the units' profiles and how they are laid.

Double-lap coverings

Plain tiles, slates, stone 'slates' and wooden shingles are all double-lap coverings. They are basically flat – with the exception of plain tiles, which have a slight camber and nibs – and are laid with their side edges butting together, not overlapping. To prevent water penetrating the joints each course is lapped in part by the two courses above it. The joints are staggered, or 'broken jointed', on alternate courses like courses of bricks in a wall.

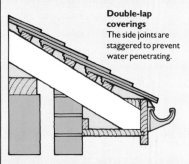

Double-lap coverings
The side joints are staggered to prevent water penetrating.

Single-lap coverings

Nearly all tiles of moulded clay and concrete are single-lap coverings, which means that each tile is profiled so as to interlock with the next one by means of a single lap on its side, and each course is laid with only a single lap at the head.

Early single-lap examples, such as clay pantiles, simply used the curved shape of the tile to form the overlap, but modern machine-made tiles of clay or concrete incorporate systems of grooves and water bars which prevent water penetrating the lap. These moulded tiles have nibs at their top back edges which hook on to the battens. They may stay in place by their own weight or be fixed with nails or clips.

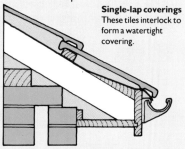

Single-lap coverings
These tiles interlock to form a watertight covering.

Tile clip

Nailed tile clip

Eave clip (flat)

Eave clip (contoured)

Verge clip (flat)

Copper rivet

GENERAL CONSTRUCTION

If you intend to make repairs yourself you will need some knowledge of the roof-covering system used on a common pitched roof. It will also help you if you have to commission contractors, either for repairs or re-roofing work, as you will benefit from a better understanding of the work being done.

Underlay

To comply with current building standards new and re-covered pitched roofs must have a weather-resistant underlay of some kind.

This underlay, sometimes called sarking, should be a reinforced bituminous felt, Type IF, or a suitable tear-resistant plastic material like polythene, which are sold in rolls to be cut to length as required.

The sheet material provides a barrier to any moisture that may penetrate the outer covering. It also improves the insulation value of the roof.

Like the tiles themselves, the sarking is laid horizontally, working upwards from the eaves, each strip overlapped by the one above it.

Battens

The roof covering is supported on sawn softwood battens which are nailed across the rafters, over the sarking (**1**). They are pre-treated with a preservative. When the roof is close-boarded there should be vertical counter-battens under the horizontal battens (**2**) to provide some ventilation under the tiles and allow any moisture to drain down freely.

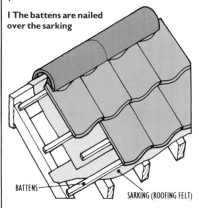

1 The battens are nailed over the sarking

BATTENS
SARKING (ROOFING FELT)

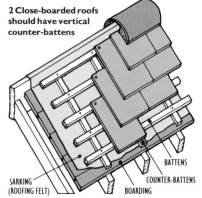

2 Close-boarded roofs should have vertical counter-battens

SARKING (ROOFING FELT)
BATTENS
COUNTER-BATTENS
BOARDING

Fixings

Most roof coverings are fixed with nails or clips. Slates and shingles are fixed individually with nails, normally two placed halfway up, though some are nailed at the top. Asbestos cement slates are centre-nailed with copper rivets to hold down the tails (See left).

Tiles have nibs that hook over the battens, keeping them in place, and some types of tile need nothing more, Those that do may be held with nails or clips.

The fixings are determined by the type and size of tile, the roof's pitch and the building's exposure.

TYPE OF ROOF COVERING						
●Black dot denotes that nail type and roof covering are compatible	Slate	Asbestos cement slate	Clay tiles	Concrete tiles	Shingles	Felt
TYPE OF NAIL						
COPPER	●	●			●	
ALUMINIUM ALLOY	●		●	●	●	
SILICON BRONZE	●	●	●	●	●	
GALVANIZED IRON					●	●
GALVANIZED STEEL					●	●

ROOF SAFETY

Working on a roof can be hazardous, and if you are unsure of yourself on heights you should call in a contractor to do the work. If you do decide to do it yourself, do not use ladders alone for roof work. Hire a sectional scaffold tower and scaffold board to provide a safe working platform complete with toe boards (▷).

Roof coverings are fairly fragile and should not be walked on. Hire crawl boards or special roof ladders to gain access. A roof ladder should reach from the scaffold tower to the roof's ridge and hook over it. Wheel the ladder up the slope and then turn it over to engage the hook (1).

Roof ladders are made with rails that keep the treads clear of the roof surface and spread the load (2), but if you think it necessary you can place additional padding of paper-stuffed or sand-filled sacks to help spread the load further.

Never leave tools on the roof when they are not being used, and keep those that are needed safely contained inside the ladder framework.

1 Engage the hook of the ladder over the ridge

2 Roof ladders spread the load

GENERAL CONSTRUCTION

It is important for the overall appearance and performance of the roof that the covering is well finished at the verge, eaves and valley edges.

Verges

In the interest of neatness the verge is normally formed by first laying an undercloak of plain tile or slate bedded on the brickwork or – in the case of an overhanging verge – nailed to the timber frame. The roof covering is then bedded in mortar on top of the undercloak and finished flush. The verge of a slate or plain tiled roof is set to slope inwards slightly to prevent rainwater running down the walls, but single lap tiles are laid flat.

Special dry-fixed verge tiles and metal extrusions are available for use with single-lap concrete tiles.

Eaves

The detail of the roof covering at the eaves depends on the type of covering. Plain tiles should have a course of under-tiles, nailed to a batten and projecting 38 to 50mm (1½ to 2in) over the gutter. The first course of tiles is laid with staggered joints over the under-tiles with their tail edges flush (1).

Some single-lap tiles, such as pantiles, also use an undercloak of plain tiles, the first course being bedded in mortar that fills up the hollow rolls. As an alternative there are also special eaves tiles with blocked ends or overhangs.

Single-lap low-profile tiles are normally laid directly over and supported by the fascia board (2).

When the roof covering is natural slate a double course is laid at the eaves. A course of short slates is nailed to the first batten and covered by a course of full slates with their tails flush and their joints staggered.

Three courses are used for asbestos cement slates to support the tail rivets used with this type of covering.

Valleys

Traditional double-lap roof coverings of plain tiles may embody special valley tiles, or, as with slate, may be formed into 'swept' or 'laced' valleys. The latter call for great skill and are expensive to make. Most valleys are formed as open gutters using sheet metal (▷).

Single-lap roofing may use sheet valleys or special trough units.

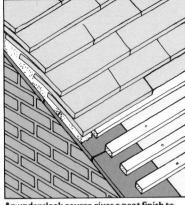

An undercloak course gives a neat finish to the verge

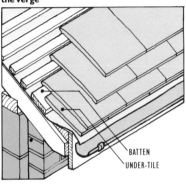

BATTEN
UNDER-TILE

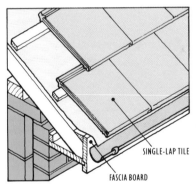

SINGLE-LAP TILE
FASCIA BOARD

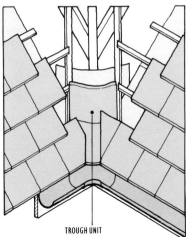

TROUGH UNIT

Verge detail at ridge
The ridge tile is set flush with the verge and filled with bedding mortar

1 Plain tiles
The eaves under-tiles are nailed to a batten. The joints between them are covered with full tiles.

2 Single-lap low-profile
Not all interlocking tiles need an under-tile at the eaves but the fascia must support the eaves course at the correct angle.

Valley tiles
Modern roofs may have trough units instead of the traditional lead sheeting.

ROOF MAINTENANCE

The roof and upper parts of a building, such as chimneys and parapet walls, must be kept in sound condition if they are to remain weatherproof. Failure of the roof covering can cause an expensive deterioration of the underlying timber structure, the interior plaster fabric and the decorative finishes.

All roof coverings have a limited life, *the length depending on the quality of materials used, the workmanship and the exposure of the house. An average roof covering might be expected to give good service for 40 to 60 years, and some materials can last for a hundred years or more, though some deterioration of the fixings and the flashings may take place. Reuse the old materials if you can.*

Patch repairs may prove to be of only temporary value and can look unsightly. If they become a recurrent chore it is time for the roof to be re-covered. This will mean stripping off the original old material and possibly reusing it, or perhaps replacing it with a new covering similar to the old.

Major roof work is not something you should tackle yourself. A contractor will do it more quickly and will guarantee the work.

Reroofing work may qualify for a discretionary improvement grant from your local authority, depending on the age and rateable value of the house. You will not require planning approval unless you live in a listed building or a conservation area.

Inspecting the roof

The roofs of older houses are likely to show their age and should be checked at least once a year.

Start by taking a general look at the whole roof from ground level. Slipped or disjointed tiles or slates should be easily spotted against the regular lines of the undisturbed covering. The colour of any newly exposed and unweathered slate will also pinpoint a fault. Look at the ridge against the sky to check for misalignment and gaps in the mortar jointing. Follow this with a closer inspection through binoculars, checking the state of the flashings at abutments and around the chimney brickwork (◁).

From inside an unlined roof you can easily spot chinks of daylight that indicate breaks in the covering, and with a light you can inspect the roof timbers for water stains, which may show as dark or white streaks. Trace the stain to find the source.

Removing and replacing a slate

A slate may slip out of place because of its nails becoming corroded or because of a breakdown of the material of the slate itself. Whatever the cause, slipped or broken slates must be replaced as soon as possible.

Use a slater's ripper to remove the trapped part of a broken slate. Slip the tool under the slate and locate its hooked end over one of the fixing nails **(1)**, then pull down hard on the tool to extract or cut through the nail. Remove the second nail in the same way. Even where an aged slate has already slipped out completely you may have to remove the nails in the same way to allow the replacement slate to be slipped in.

You will not be able to nail the new slate in place. Instead cut a 25mm (1in) wide strip of lead or zinc to the length of the slate lap plus 25mm (1in) and nail the strip to the batten, nailing between the slates of the lower course **(2)**. Then slide the new slate into position and turn back the end of the lead strip to secure it **(3)**.

1 Pull out nails

2 Nail strip to batten

3 Fold strip over edge

Cutting slate

Cut from each edge

With a sharp point mark out the right size on the back of the slate, either by measuring it out or scribing round another slate of that size. Place the slate, bevelled side down, on a bench, the cutting line level with the bench's edge, then chop the slate with the edge of a bricklayer's trowel. Work from both edges towards the middle, using the edge of the bench as a guide. Mark the nail holes and punch them out with a nail or drill them with a bit the size of the nails. Support the slate well while making the holes.

Asbestos cement slates

These can be cut by scribing the lines, then breaking the slates over a straight edge or sawing with a general-purpose saw. If you saw them wear a mask, keep dust damped down well and sweep it into a plastic bag for disposal.

REPLACING A TILE

Individual tiles can be difficult to remove for two reasons: the retaining nibs on their back edges and their interlocking shape which holds them together.

You can remove a broken plain tile by simply lifting it so that the nibs clear the batten on which they rest, then drawing it out. This is made easier if the overlapping tiles are first lifted with wooden wedges inserted at both sides of the tile to be removed **(1)**.

If the tile is also nailed try rocking it loose. If this fails you will have to break it out carefully. You may then have to use a slater's ripper to extract or cut any remaining nails.

Use a similar technique for single-lap interlocking tiles, but in this case you will also have to wedge up the tile to the left of the one being removed **(2)**. If the tile is of a deep profile you will have to ease up a number of the surrounding tiles to get the required clearance.

If you are taking out a tile to put in a roof ventilator unit you can afford to smash it with a hammer. But take care not to damage any of the adjacent tiles. The remaining tiles should be easier to remove once the first is out.

1 Lift the overlapping tiles with wedges

2 Lift interlocking tiles above and to the left

CUTTING TILES

To cut tiles use an abrasive cutting disc in a power saw or hire an angle grinder for the purpose. Always wear protective goggles and a mask when cutting with a power tool.

For small work use a tungsten grit blade in a hacksaw frame or, if trimming only, pincers but score the cutting line first with a tile cutter.

REBEDDING RIDGE TILES

Ridge tiles on old roofs often become loose because of a breakdown of the old lime mortar.

To rebed ridge tiles first lift them off and clear all the old crumbling mortar from the roof, and from the undersides of the tiles.

Give the tiles a good soaking in water before starting to fix them. Mix a new bedding mortar of 1 part cement to 3 parts sand. It should be a stiff mix and not at all runny. Load about half a bucketful and carry it on to the roof.

Dampen the top courses of the roof tiles and throw the mortar from the trowel to form a continuous edge bedding about 50mm (2in) wide and 75mm (3in) high, following the line left behind by the old mortar.

Where the ridge tiles butt together, or come against a wall, place a solid bedding of mortar, inserting pieces of slate in it to reduce shrinkage. Place the mortar for all the tiles in turn, setting each tile into place and pressing it firmly into the mortar. Strike off any squeezed-out mortar cleanly with the trowel, without smearing the tile. Ridge tiles should not be pointed.

Apply bands of bedding mortar on each side

Insert pieces of slate in joint bedding mortar

HALF-ROUND   HOG-BACK   ANGLE

Typical ridge tile shapes

WEAR A GAUZE MASK WHEN WORKING WITH ASBESTOS CEMENT SHEET

SHEET ROOFING

The commonest sheeting materials for roofing are corrugated asbestos cement and rigid PVC plastic. Aluminium and steel corrugated sheeting are made for roofing but are not generally used for domestic work.

Sheet roofings are used mainly for outbuildings such as garages and garden sheds, and the plastic types may be used for lean-to extensions.

Consult your local Building Control Officer when considering a plastic roofing to ensure that it complies with the fire regulations.

Corrugated sheet roofing

Corrugated sheeting is produced in standard profiles of 32mm (1¼in) for plastic, and 75mm (3in) and 150mm (6in) for plastic and asbestos cement. When calculating the number of sheets you need you must make an allowance for the side overlap. Small-profile sheeting should overlap at least two corrugations (1), while larger ones can be used with only one (2). The end overlap should be at least 150mm (6in) in sheltered locations for roofs with pitches of about 22 degrees or more. For pitches of less than this a 300mm (1ft) overlap should be allowed.

Corrugated sheeting is carried by purlins (▷). They are of wood in most domestic buildings, though some system-built garages embody steel sections. Wood screws or drive screws fix the roofing to wooden purlins (3) and hooked bolts are used with metal ones (4). There are special plastic washers and caps for sealing the heads of the screws or bolts.

Cutting corrugated plastic
On thin plastic sheet mark the cutting lines with a felt-tipped pen, then cut it with a fine-toothed tenon saw. Support the sheeting between two boards on trestles and use the top board as a guide for your saw.

When cutting to length saw across the peaks of the corrugations with the saw held at a very shallow angle. Cut halfway through, then turn the sheet over and cut from the other side.

When cutting to width make the cut along the peak of a corrugation, again working with the saw at a shallow angle. For a cut near the middle of the sheet support the sheet on both sides of the cutting line from below.

Cutting corrugated asbestos
Lay the sheet on boards supported by trestles. As the material is fairly thick you will not need a top supporting board and you can saw the sheet without turning it over. Use a sheet saw or general-purpose saw, damping down the dust and sweeping it into a plastic bag, to be sealed with adhesive tape and put in a dustbin. Damp newspaper under the trestles will help contain the dust.

Laying corrugated sheet

You can work from stepladders inside the structure or from above on boards, depending on the pitch of the roof.

Start at the eaves and work from left to right or vice versa.

Asbestos cement sheet is laid smooth side up.

Plastic sheet
Position the first sheet and drill oversized clearance holes for the fixing screws on the centre lines of the wooden purlins. Drill the holes in the crowns of the corrugations and space them at about every third or fourth corrugation. Never place fixings in the troughs of the sheeting.

Don't make holes where the next sheet will overlap the first. Instead lay the second sheet with two corrugations overlapping and drill through both sheets. Lay and fix the rest of the row of sheets in the same way.

Start the next row on the same side as the first one (1), overlapping the ends by at least 150mm (6in), and drill and fix through both layers.

Finally fit the protective plastic caps over the fixing washers.

Laying asbestos sheet
Corrugated asbestos sheet is fairly thick material, and to deal with the bulk of four thickness at the end laps you should cut some of the corners away to form mitres (2). The angle of the mitre is drawn between two points representing the length of the end lap and the width of the side lap.

Fix the sheets as for plastic sheet (See above). For the large profile sheeting use only two fixings to each purlin, placed adjacent to the side laps. The end laps are fixed through both layers.

SEE ALSO

Details for: ▷

Trussed roof	43
Roof safety	47
Sealing roofs	55

1 Small-profile lap

2 Large-profile lap

3 Wood purlin fixing

4 Metal purlin fixing

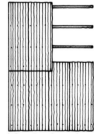

1 Overlap the ends

2 Mitre the corners

49

FLAT ROOFS

Timber-framed flat roofs are used for the main roof and rear extensions and for outbuildings. Most have joists carrying stiff wooden decking, and these usually cross the shorter span, spaced at 400mm (1ft 4in), 450mm (1ft 6in) or 600mm (2ft) between centres. Herringbone or solid strutting is needed for a span of more than 2.5m (8ft) to prevent the joists buckling. Their ends may be fixed to wall plates on load-bearing walls or, as on an extension, they are set in metal hangers or into the brickwork of the wall. Metal restraint straps tie the ends of the timbers down to the walls.

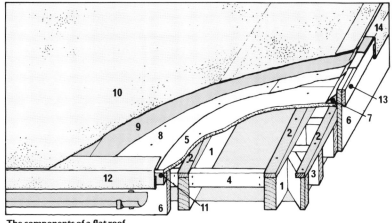

The components of a flat roof

1 Joists	5 Decking	9 Second felt layer	13 Verge drip batten
2 Furring	6 Fascia board	10 Third felt layer	14 Felt verge drip
3 Return-joist	7 Angle fillet	11 Eaves drip batten	
4 Nogging	8 First felt layer	12 Felt eaves drip	

Fall

The fall of a flat roof should be at least 1:80 for smooth surfaces like metal or plastic, and 1:60 for rougher materials. It is meant to shed water, but if too shallow will allow puddles to form. These, aided by sun or frost, can break some roof coverings down and let the standing water get through.

To give the fall the joists may be sloped, but this means that any ceiling will also slope. Where a flat ceiling is wanted tapered 'furrings' are nailed to the tops of the joists. Otherwise joists may be set across the line of the fall with parallel furring pieces of decreasing thickness nailed to them or tapered furrings fixed across them. The latter gives better cross ventilation.

Furring methods

1 Tapered furrings fixed in line with joists

2 Tapered furrings fixed across joists

3 Furrings of decreasing size fitted across fall

Decking

A decking in the form of boards – square-edged or tongued and grooved – or of such sheet materials as plywood or chipboard, is fixed to the joists to make a flat support for the final covering.

Solid boarding of the kind found in older flat roofs should be laid with its joints running with the fall for most efficient water-shedding.

Plywood and chipboard, normally 18mm (¾in) thick, are laid with their long edges across the joists and their ends centred over supporting joists. Noggings may be fitted between the joists to give extra support to the long edges of the board, depending on its thickness and the joist spacing.

If you wish to have a felted roof you might consider using the pre-felted chipboard now available. This material is laid with 3mm (⅛in) gaps between the boards to allow for thermal expansion, and fixed down with nails or screws. The gaps between the boards are filled with a cold bonding mastic and then sealed with tape.

Pre-felted chipboard provides a weatherproof decking that does not need the built-up roof covering to be applied immediately as do others.

Finishing

The decking is waterproofed with either mastic asphalt or roofing felt (See right). In addition, a felted roof may be covered with a layer of 12mm (½in) light-coloured chippings which reflect some of the sun's heat from the roof.

FLAT ROOF COVERINGS

Bitumen-based coverings fall into two types: asphalt and bituminous felt. They are much better than they once were and are now generally used on domestic buildings instead of the expensive lead, zinc or copper seen on older houses.

MASTIC ASPHALT

This waterproof material, made from either natural or synthetic bitumen, weathers very well. It is melted in a cauldron and spread over the roof hot, to set in an impervious layer. Two layers are applied with a float to a combined thickness of 18mm (¾in) on a separating layer of sheathing felt that covers the decking. The felt is loose-laid with 50mm (2in) lapped joints. It allows movement in the substrate without affecting the asphalt. If the substrate has a pre-felted or other surface likely to adhere, a non-bituminous building paper is placed beneath the felt layer.

Laying hot asphalt is a skilled job.

ROOFING FELTS

These bitumen-impregnated sheet materials are applied in layers to produce 'built-up' roofing, bonded with hot or cold bitumen. Making such a roof with hot bitumen is a professional job.

Several felts are available, and the choice can affect a roof's cost and its long-term performance. Traditional British Standard felts are classified by their reinforcing base material and finish, indicated by a number and letter. A colour strip identifies the base material (See opposite). These cheaper felts are well tried but less tough than modern high-performance ones. The latter are based on glass tissue reinforced with polyester or polyester fabric, some with modified bitumen for greater flexibility.

BONDING FELTS

Plywood, chipboard, wood-wool and concrete decks need partial bonding of the first layer, using a perforated underlay either applied loose or bonded with bands of bitumen about 500mm (1ft 8in) wide around the edges. The first layer is then only partially bonded by the hot bitumen penetrating through the holes. On solid timber the first layer is secured with clout nails. Subsequent layers are fully bonded by applying hot or cold bitumen with a trowel or notched spreader over the whole surface.

ABUTMENTS AND PARAPETS

Wherever a flat roof abuts a house or parapet wall leaks can occur, so the covering is usually turned up the wall – called a 'skirting' – and tucked into the mortar bed of the brickwork, or covered by flashing (▷).

Parapet walls are prone to damp, being exposed on both sides. Their top edges are usually finished with brick, stone or tile coping which should overhang the wall faces to throw off rainwater. Damp-proof courses of lead, bituminous felt or asphalt must underlie copings (**1**).

Parapet walls of no more than 350mm (1ft 2in) may have the skirting taken up their face and continued under the full width of the coping (**2**).

When a parapet wall's inside face has no impervious layer a second damp-proof course is needed directly above the skirting. A solid wall may have asphalt roof covering taken up two courses of bricks, then continued across the wall to form a DPC (**3**). Or a flexible DPC like bituminous felt or lead may be set in the bed joint before the roofing is laid and dressed down to form a flashing over the skirting (**4**).

Cavity parapet walls also need a DPC where the roof abuts. Water penetrating from the roof side must be prevented from running down the cavity to damage the interior walls below roof level (**5**). The DPC is stepped up from the inside leaf across the cavity to the outer leaf to form a cavity tray. Cement rendering is not a satisfactory solution as the inevitable movement of the wall can cause cracks that will let in water.

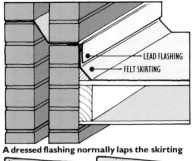
LEAD FLASHING
FELT SKIRTING

A dressed flashing normally laps the skirting

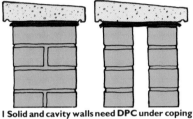

1 Solid and cavity walls need DPC under coping

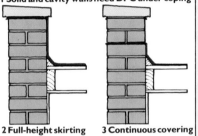

2 Full-height skirting **3 Continuous covering**

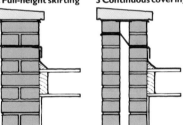

4 DPC flashing **5 DPC cavity tray**

FITTING A NEW CAVITY TRAY

A cavity wall abutted by an extension roof needs a cavity tray to protect it from damp. Normally it would be built-in, but in existing buildings with new extensions special trays – moulded units of polypropylene – can be inserted from outside by removing a course of bricks. A tray equals two brick-lengths and can also be used singly where a cavity is bridged by an extractor.

Inserting the tray

Two courses above the proposed roof level remove three bricks, whole if possible, and without letting rubble fall into the cavity. On the cleaned bricks lay a length of flashing wide enough to project 50mm (2in) into the wall and cover the roof skirting by 75mm (3in) when dressed down. Trap the flashing with the first tray unit, pushing it into one end of the opening (**1**). Place two bricks in the tray and bed and joint them in mortar (**2**). Pack out the top joint with slate pieces and fill it with mortar pushed well home but not out at the back. Rake out a weep hole at the base of the middle joint with a wire.

Cut out two more bricks, again leaving a three-brick opening (**3**), roll out the flashing and insert a second tray. Join the trays with the clip provided, fitting it over the meeting ends to make a water tight joint (**4**), and lay two more bricks in the opening. Continue in this way until the tray is the required length. Only one brick need be removed at the end to make a two-brick opening for the last unit.

SEE ALSO

Details for: ▷
Flashings   54–55
Ventilators   70, 73, 77

Moulded cavity tray
Straight and angled sections are available from most builders' merchants.

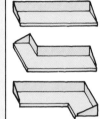

1 Trap the flashing

2 Lay two bricks

3 Cut out two bricks

4 Insert second tray

TYPES OF FELTS FOR FLAT ROOFS

Felt type British Standard Ref.	Base	Surface Finish	Colour code	Weight Kg/per roll	Properties and uses
BS 747 **1B**	Fibre	Sand	White	36 Kg (79lb)	Least expensive type. Relatively weak. Good for roofing outbuildings.
BS 747 **1E**	Fibre	Mineral	White	38 Kg (84lb)	
BS 747 **2B**	Asbestos	Sand	Green	36 Kg (79lb)	Good fire-resistance. More expensive than glass-fibre base felts. Can be nailed on sloping roofs. Use for underlays, vapour checks or complete systems.
BS 747 **2E**	Asbestos	Mineral	Green	38 Kg (84lb)	
BS 747 **3B**	Glass fibre	Sand	Red	36 Kg (79lb)	Rot proof, inexpensive, unsuitable for nailing. Good for 2 or 3 layer systems.
BS 747 **3E**	Glass fibre	Mineral	Red	28 Kg (62lb)	
BS 747 **3G**	Glass fibre	Grit underside Sand topside	Red	32 Kg (70lb)	Perforated first layer for partial bonding systems.
HIGH PERFORMANCE FELTS					
NO BS NUMBERS	Glass/polyester	Sand	Black	36 Kg (79lb)	Rot proof, tough, good weathering, can be nailed. Use for 2 or 3 layer systems.
	Glass/Polyester	Mineral	Blue/grey. Green	28 Kg (62lb)	
BS 747 **5U, 5B** BS 747 **5E**	Polyester	Sand	Blue	18,42Kg (40,93lb)	More expensive than glass polyester but better performance. Use for 2 or 3 layer systems. The best type for house extensions.
	Polyester	Mineral	Blue	47Kg (104lb)	

RENEWING A FELT ROOF

Covering flat roofs, or stripping and re-covering them, should usually be left to professionals. A built-up felt system using hot bitumen or torching — using a gas-powered torch to soften bitumen coated felt — is beyond the amateur.

Yet a competent person can confidently replace perished felt on a garage roof, using a cold bitumen adhesive. The following example assumes a detached garage with a solid timber decking covered with three layers of felt.

Built-up felt system
Lap the edges of the felt strips and stagger the joints in alternate layers.

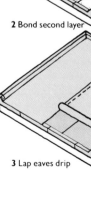

1 Nail first layer

2 Bond second layer

3 Lap eaves drip

Replacing perished felt

Wait for dry weather, then strip the old felt. Pull out any clout nails and check the deck for distorted or rotten boards. Lift and replace unsound ones with new ones, using galvanized wire nails and punching them below the surface.

Cutting to stagger joints

For a three-layer build-up start at one edge with a strip about one third of the roll width. The second layer starts with two-thirds width, the top layer with a full width. A two-layer roof starts with half a roll width, then with a full one. You can modify this to suit your roof and avoid a too-narrow strip at the other edge. If, for economy or ease of handling, you use short lengths the end should overlap at least 100mm (4in), the lower piece always lapped by the higher one as you work up the slope.

First layer

Cut the strips for the first felt layer slightly longer than the slope of the roof, and allow for an overlap of at least 50mm (2in) at the long edges. Cut one narrow side length so that successive lapped joints will be staggered.

Nail the felt down with 18mm (¾in) clout nails 50mm (2in) apart down the centres of the laps and 150mm (6in) apart overall (1). Tuck and trim the felt into the corners of the verge upstand to get mitred butt joints. Trim the felt

flush at eaves and verges and form and fit the drip at the eaves (See right).

Second layer

Cut lengths for the second layer, put the side piece in place, then roll it back halfway from the eaves end. Brush or trowel cold bitumen adhesive on the felt below but not on the verge upstand. Re-lay the felt, press it down, then roll up the other half and repeat, ensuring that the adhesive is continuous (2).

Fold and tuck the felt into the corner of the verge upstand and trim it to a mitred butt joint. Turn it back from the verge, apply adhesive and press it into place against the upstand. Trim the end to butt against the edge of felt for the eaves drip (See below) and trim the other edges flush with the verge. Place the next length, overlapping the first by at least 50mm (2in), and again roll back each half in turn, applying the adhesive. Repeat this across the roof, and cut and tuck the felt at the other verge corner.

Third layer

Cut and lay the mineral felt top layer or 'capsheet' in the same way as the second layer but lap the eaves drip and not the verge upstands (3).

Cut strips of the felt to form verge drips (See right) and nail and bond them into place around the side and near edges.

MAINTAINING A FLAT ROOF

Whatever the material used for covering your flat roof it is sensible to carry out a routine inspection of its current condition at least once a year.

Climb on to the roof to inspect it, wearing soft-soled shoes, and make the check in two parts: once on a dry day and again shortly after there has been rain, when standing water can be seen.

Remove any old leaves, twigs or other matter that may have found its way there and brush off any silt deposits that may have built up. Note the positions of puddles because, although

they may not present an immediate problem, leaks occurring later in the life of the roof will be more easily traced. In an asphalt or felted roof you should also note any blisters or ripples.

On a felted roof test the overlapping joints to see that they are still well bonded. Also make a close inspection of the vulnerable edges of the roof. Check the soundness of the coving at the verges and eaves and the flashings at abutments. You should also inspect the condition of the gutterings and outlets and remove any blockages.

MAKING DRIPS

Eaves

Cut 1m (3ft 3in) long strips from the length of a roll of felt. Calculate the width of the strips by measuring the depth of the drip batten and adding 25mm (1in), then doubling this figure and adding at least 100mm (4in). Cut away 50mm (2in) from one corner to enable the ends to be overlapped and make folds in the strips, using a straight edge. Nail the drip sections to the drip batten with galvanized clout nails and fold each strip back on itself and bond it over the first layer of felt (1).

Cutting the corners

Where the drip meets the verge, cut the corners to cover the end of the upstand (2). If necessary, make a paper pattern before cutting the felt. You may need an extra wide strip to allow for a tall upstand. Fold the tabs and bond into place, except the end one, which is left free to be tucked into the verge drip.

Verge drips

Cut and fix the verge drips in place after the top layer of roofing. Cut the strips 1m (3ft 3in) long and calculate their width as with the eaves drip, but allow extra for the top edge and slope of the upstand. Working from the eaves, notch the ends of the strips where they overlap, as described for eaves.

Cut and fold the end of the first strip where it meets the eaves (3), nail the strip to the batten and bond the remainder into place (4).

At the rear corners cut and fold the end strip covering the side verge (5). Cover the rear verge last, cutting and folding the corners to lap the side pieces and finishing with neat mitres (6).

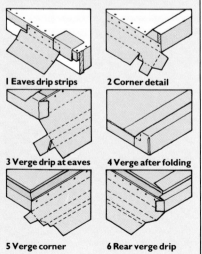

1 Eaves drip strips   2 Corner detail

3 Verge drip at eaves   4 Verge after folding

5 Verge corner   6 Rear verge drip

FAULT-FINDING

Damp patches
Damp patches on a ceiling are a clear sign that the roof needs attention, but the source of the problem is not always obvious. If they are near an internal wall you should suspect a breakdown of the flashing details.

Locating the leak
A leak anywhere else in the roof may be hard to find, as the water can run downhill from its entry point before dripping on to the ceiling. Measure the distance between the patch and the edges of the ceiling, then locate the point on the roof surface and work from it up the slope seeking the source.

Splits and blisters
Splits and blisters on the smooth surface of an asphalt or bitumen felt covering may be obvious, but chippings on a covering can be troublesome and must be cleared away. This is not easy, and the attempt can obliterate the cause of the leak. Use a blowtorch or hot-air paint stripper to soften the bitumen and scrape the chippings away. The surface must be made smooth if it is to be patch-repaired.

Splits in the covering caused by some movement of the substrate can be recognised by the lines they follow. Blisters formed by trapped moisture or air should be pressed to locate any weaknesses in the covering, which will show as moisture is expelled. These must be sealed with patches. Their cause may be moisture permeating the substrate from below and, heated by the sun, expanding under the covering. If this is undamaged the blister can be left, but its cause must be dealt with.

Damp and condensation
Damp near a wall may be caused by porous brickwork above the flat roof, lack of pointing or DPC, slipped or inadequate coping on parapet walls and/or a breakdown of flashings, which should be made good as required. Condensation may also cause dampness, and is potentially a more serious problem. If warm moist air permeates the ceiling the vapour can condense under the cold roof and start rot in the structural timbers. In such a case upgrade the ceiling with a moisture check and fit some type of ventilation (▷). If the problem is not solved so easily, have the roof recovered and also include better insulation.

FLAT ROOF REPAIRS

Just how a flat roof is to be repaired will depend on its general condition and age, and the extent of the damage. If inspection shows that the surface of the covering has decayed, as may happen to some traditional bitumen felts, it may be better to call in a contractor and have the roof re-covered.

Patch repairs

Such localised damage as splits and blisters can be patch-repaired with the aid of proprietary repair kits, but, as their effectiveness is only as good as their adhesion to the background, take care when cleaning the surface. Kill any lichen or moss spores with a good fungicide or bleach before starting the repair work.

A patched roof, if visible from above, can be rather an eyesore. This can be corrected with a finishing coat of bitumen and chippings or reflective paint to unify the surface area. Work on a warm day, preferably after a spell of dry weather.

Dealing with splits

You can use most self-adhesive repair tapes to patch-repair splits in all types of roof coverings.

First remove any chippings (See left), then clean the split and its surrounding surface thoroughly. Fill a wide split with a mastic compound before taping. Apply the primer supplied over the area to be covered and leave it for an hour.

Where a short split has occurred along a joint in the board substrate prepare the whole line of the joint for covering with tape.

Peel back the protective backing of the tape and apply it to the primed surface(1). If you are working on short splits cut the tape to length first. Otherwise work from the roll, unrolling the tape as you work along the repair.

Press it down firmly and, holding it in place with your foot, roll it out and tread it into place as you go, then cut it off at the end of the run. Go back and ensure that the edges are sealed (2).

1 Apply the tape

2 Press tape firmly

Dealing with blisters

Any blisters in asphalt or felted roofs should be left alone unless they have caused the covering to leak or they contain water.

To repair a blister in an asphalt roof first heat the area with a blow torch or hot-air stripper and, when the asphalt is soft, try to press the blister flat with a block of wood. If water is present cut into the asphalt to open the blister up and let the moisture dry out. This can be encouraged by careful use of the heat before pressing the asphalt back into place. Work mastic into the opening before closing it, then cover the repair with a patch of repair tape.

In a blister on a felted roof, make two crossed cuts and peel back the covering. Heating the felt will make this easier. Dry and clean out the opening, apply bitumen adhesive, and when it is tacky nail the covering back into place with galvanized clout nails (3).

Cover the repair with a patch of roofing felt, bonded on with the bitumen adhesive. Cut the patch so as to lap at least 75mm (3in) all round. Alternatively you can use repair tape.

Treating the whole surface

A roof which has already been patch-repaired and is showing general signs of wear and tear can be given an extra lease of life by means of a liquid waterproofing treatment.

The treatment consists of a thick layer of cold-applied bitumen and latex rubber waterproofer which can also be reinforced with an open-weave glass fibre membrane.

First sweep the roof free of all dirt and loose material and treat the surface with a fungicide to kill off any traces of lichen and moss.

Following the maker's instructions, apply the first coat of waterproofer with a brush or broom (4), then lay the glass fibre fabric into the wet material and stipple it with a loaded brush. Overlap the edges of the fabric strips by at least 50mm (2in) and bed them down well with the waterproofer. Clean the brush with soapy water.

Let the first coat dry thoroughly before laying the second and allow that one to dry before applying the third and last coat. When the last coat becomes tacky cover it with fine chippings or clean sharp sand to provide it with a protective layer.

SEE ALSO
Details for: ▷

Ventilation	71
Organic growth	32
Parapets	51
Flashings	54–55
Condensation	10–11

3 Nail cut edges
Use bitumen adhesive to glue a felt patch over the repair.

4 Brush on first coat

FLASHINGS

Flashings are used to weatherproof the junctions between a roof and the other parts of the building, which are usually at the abutments with walls and chimneys and where one roof meets another.

Where flashing is used
Typical types of flashing for pitched and flat roofs ▶
1 Valley
2 Apron
3 Wall abutment
4 Parapet abutment
5 Chimney abutment

Flashing materials

The most common flashing materials are lead, zinc, roofing felt and mortar fillets. Of all these lead is by far the best because it weathers well, is easily worked – though shaping it is generally a craft skill – and can be applied to any situation and roof covering.

Zinc is a cheaper substitute for lead and is not so long-lasting or so easy to work into shape. Where zinc flashings need to be replaced the extra cost of using lead would be easily repaid because of lead's much longer life.

Bitumen felt may be used for flashings on felted roofs, but this material cannot be easily manipulated and is used normally for the more simple cover flashings that overlap the upturned skirtings of the felt roofing.

Flashings of mortar fillets are common on the pitched roofs of older terraced houses, but are prone to shrinkage. They are still used for reasons of economy or for re-roofing work, sometimes with inset cut tiles.

Flashing construction

The design of a given flashing will be determined by the particular details at the junction and to some extent by the materials being used.

Typical situations are generally treated in a standard way, and these are shown here, using lead for the flashings.

Abutments
Where the inclined edge of a typical pitched roof abuts a wall the type of flashing used is determined by the nature of the roof covering and the pitch of the roof.

Double-lap tiles
Slate or plain tiled roofs with a pitch of 30 degrees or more, normally use soakers and a cover flashing. Soakers are lead or zinc pieces, equal in length to the tile's gauge plus the lap, folded into right angles lengthwise. The part that lies on the tiles should be at least 100mm (4in) wide and the upstand 75mm (3in). The back edge turns down over the tile's top edge, so 12 to 25mm (½ to 1in) extra in length is added.

A soaker is laid over the end tile or slate as a course is laid, the upstands flat against the brickwork and protected by a stepped cover flashing dressed down over them. The flashing's top edges are turned into the bed joints, held by lead wedges and pointed with mortar (**1**).

Single-lap tiles
Contoured single-lap tiles can be treated at abutments with one-piece flashing. The lead is tucked into the brick wall by the stepped or chased method and dressed down over the tile, the amount of overlap depending on tile contour and roof pitch. On a shallow pitch it should be at least 150mm (6in). The lead is dressed to the tile's shape and the step at each course and its free edge carried over the nearest raised tile contour (**2**).

Valley flashing
Some plain tiled roofs have special valley tiles that take the tiling into the angle, but most tiled and slated roofs have valley gutters of metal flashing. A valley flashing is made by laying a lead lining on boarding that runs from eaves to ridge, following the angle of the valley, and dressing it over wooden fillets nailed to the boarding to form an upstand (**3**).

Where two valleys meet at the ridge a lead saddle is formed. The edges of the tiles or slates are cut to follow the angle of the valley and to leave a gap of no less than 100mm (4in) between them. Slate coverings should overhang the supporting valley fillet by 50mm (2in) and contoured tiles should be bedded in mortar and finished flush with the edge of the tiles to form a watertight gutter.

1 Double-lap flashing
SOAKERS
WEDGES
FLASHING

2 Single-lap flashing
DRESSED FLASHING

3 Valley flashing
FILLET

APRON FLASHING

The head of a lean-to roof is weatherproofed with a lead apron flashing, its top edge pointed into a mortar joint two courses above the roof, dressed down on to the roof. It should lap the roof covering by 150mm (6in) or more.

Special moulded flashing units are available for use with corrugated sheet roofing of plastic or asbestos cement. These are shaped to fit the contour of the roofing and have flat hinged upstands which adapt to any roof slope. The upstand can have conventional cover flashing or can be sealed with self-adhesive flashing tape.

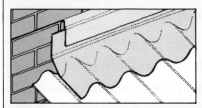

Moulded apron flashing for corrugated roofing

CHIMNEY FLASHING

The flashing where a roof meets the side of a chimney is much the same as at an abutment, but there are junctions at the front and back of the chimney.

An apron flashing is fitted in front, the upstand is returned on to the sides of the stack and its top edge set in a bed joint. The apron, extending beyond the chimney sides, is dressed to the tile contour. Side flashings are fitted, using soakers or one-piece flashing, and the back of the chimney receives a timber-supported back gutter. The lead's front edge is turned up the brick face, its ends folded on the side flashings, and a separate cover flashing is dressed over the upstand.

The back of the gutter follows the roof slope and is lapped by the tiles, which are fitted last.

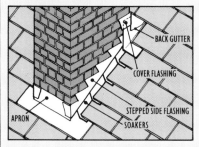

BACK GUTTER
COVER FLASHING
STEPPED SIDE FLASHING
APRON
SOAKERS

Chimney flashing for a slate roof

FLASHING REPAIRS

Many problems come about with flashings because of corrosion of the flashing material or failure of the joints between different materials due to erosion or thermal movement.

A perished flashing should be stripped out and replaced. If this will require craft skills the work should be done by a specialist contractor, but in many cases leaks may be caused by shrinkage cracks, and you can repair these with self-adhesive flashing materials.

Using caulking compound

Cement fillets often shrink away from wall abutments. If the fillets are otherwise sound – free of cracks for example – fill the gap with a gun-applied flexible caulking compound, having chosen a colour to match the fillet. Brush the surfaces to remove any loose material before injecting the compound.

Flashing tape

Prepare the surfaces by removing all loose and organic material. A broken or crumbling cement fillet should be made good with mortar.

Ensure that the surfaces are dry and if necessary apply a primer – it is supplied with some tapes – about one hour before using the tape **(1)** .

Cut the tape to length and peel away the protective backing as you press the tape into place. Work over the surface with a cloth pad, applying firm pressure to exclude any air trapped underneath the tape **(2)** .

1 Apply a primer with a 50mm (2in) paintbrush

2 Press tape with a pad to exclude air bubbles

Repointing flashing

Metal flashings which are tucked into brickwork may have worked loose where the old mortar is badly weathered. If the flashing is otherwise sound rake out the mortar joint, tuck the lead or zinc into it and wedge it there with rolled strips of lead spaced about 500mm (20in) apart. Then repoint the joint. While you have the roof ladders and scaffolding in place, rake out and repoint all of the mortar joints if in poor condition.

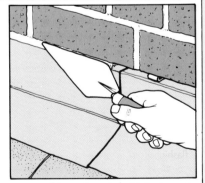

Rake out joint and repoint with fresh mortar

Patching lead

Lead will not readily corrode but splits can occur in it where it has buckled through expansion and contraction over the years.

Flashing tape can be used but it is possible to patch lead by soldering or, for a more substantial repair, cutting away a weak or damaged portion and joining on a new piece by lead 'burning' or welding. This is not a job you can easily do yourself and it should be handed over to a professional. It should be done only when it is more economical to have the old lead repaired than to have a new flashing made and fitted.

SEALING GLAZED ROOFS

Traditional timber-framed conservatories, glazed porches and greenhouses can suffer from leaks caused by a breakdown of the seal between the glass and the glazing bars. Minor leaks should be dealt with promptly because the trapped moisture could lead to timber decay and expensive repair work.

Using aluminium tape

You can waterproof glazing bars with self-adhesive aluminium tape simply cut to length and pressed in place.

Clean all old material from the glazing bars and the edges of glass on both sides of them, let the wood dry out if necessary and apply wood primer. When the primer is dry fill the rebates with putty or mastic. The width of tape must cover the upstand of the bars and lap the glass each side by 18mm (¾in).

Start at the eaves and work up the roof, peeling off its backing as you unroll the tape. Mould it to the glazing bars' contour, excluding air bubbles.

At a step in the glass cut the tape for a 50mm (2in) overlap and mould the end over the stepped edge, then start a new length lapping the stuck-down end by 50mm (2in) and so on, to the ridge.

At the ridge you may cut the tape to butt against the framework or lap on to it. A horizontal tape should cover the turned ends. Where a lean-to roof has an apron flashing tuck the tape under it.

Self-adhesive aluminium tape can be painted or left its natural colour.

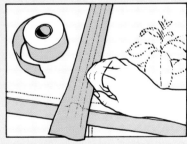

Clean the surfaces before applying the tape.

Clear tape

You can make a temporary repair to cracked glass with clear self-adhesive waterproofing tape.

Clean the glass and apply the tape over the crack on the weather side. It will make an almost invisible repair, especially if done promptly.

You can also use this tape for sealing the overlaps on translucent corrugated plastic roofing.

GUTTERING

Guttering collects the rainwater that runs down a roof and leads it to a downpipe through which it discharges into a drain. Good rainwater disposal is vital in preventing damp developing in the fabric of a house.

The guttering on domestic buildings is used in various ways adapted to the design of the roof.

Roof drainage

The size and layout of a roof drainage system should enable it to discharge all the water from a given roof area. The flow load determining the guttering's capacity depends mainly on the area of the roof. Makers of rainwater goods usually specify the maximum area for a given size and profile of gutter based on a rainfall rate of 75mm/hr (3in/hr). If you replace an old gutter, perished but of the right capacity, have the size that of the original or slightly larger.

A downpipe's position can affect the system's performance. A system with a central downpipe can serve double the roof area of one with an end outlet. A right-angled bend will reduce the flow capacity by about 20 per cent if it is near the outlet.

In practice, unless you are working on a new building, the positions of drains and downpipes are already fixed.

Sizes
Gutter sizes are generally specified by their overall width in cross section and sometimes by their depth as well.

EAVES GUTTERS

The commonest types by far are the eaves gutters, which are fixed along the eaves fascia boards (See below). They are made in many materials and designs.

PARAPET GUTTERS

Parapet gutters are generally found in older houses and may serve a flat or pitched roof set between two parapet walls. This type is generally purpose-made as part of the original structure of the roof and is usually covered with a metal or bituminous roofing material.

VALLEY GUTTERS

Valley gutters are a form of flashing used at the junctions between sloping roofs. They are not gutter systems in themselves but they direct the rainwater into eaves or parapet gutters (◁).

TYPICAL PROFILES AND SIZES OF GUTTERING

Half round	Ogee (OG)	Moulded OG	Box
75mm (3in)			
100mm (4in)	100mm (4in)	100 x 75mm (4 x 3in)	100 x 75mm (4 x 3in)
112mm (4½in)	112mm (4½in)		
125mm (5in)	125mm (5in)	125 x 100mm (5 x 4in)	125 x 100mm (5 x 4in)
150mm (6in)	150 x 100mm (6 x 4in)	150 x 100mm (6 x 4in)	

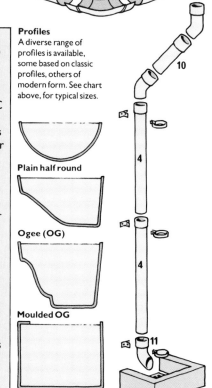

1 Stopend
Internal and external fittings for socketed or non-socketed types.

2 Gutter brackets
Normally screwed to fascia board, but some can fix to rafter bracket arms.

3 Guttering
In various profiles and lengths of 1.8 to 3m (6 to 10ft) with socket at one end or spigots both ends.

4 Downpipe
Available in 1.8 to 3m (6 to 10ft) lengths. Metal types may have integral fixing lugs.

5 Hopperhead
May be used as part of downpipe system to receive waste pipes from another source.

6 Pipe clip
Secures downpipes to wall.

7 Running outlet
May be double or single socketed.

8 Gutter angle
Available in 90, 120 and 135 degree angles in most systems for turning corners.

9 Stopend outlet
Used with downpipe at an end.

10 Offset
Used on guttering fitted to overhanging eaves. Available in standard projections or can be made up with special offset bends and a length of downpipe.

11 Shoe
Throws water clear of a wall into open gulley (◁).

EAVES GUTTER SYSTEMS

Eaves gutter systems are fabricated in cast iron, cast and rolled sheet aluminium, asbestos cement and a uPVC rigid plastic. With the exception of the roll-formed aluminium type the systems are made up from basic lengths of gutter and downpipes with a range of fittings (See diagram).

Moulded or cast gutters have a socket at one end into which the plain spigot end of the next section is jointed.

For gutters that are not symmetrical in section, such as an OG, this means that components are left- or right-handed, and this has to be noted when replacement parts are being ordered.

Traditional cast iron and modern aluminium guttering may be compatible should you wish to renew only a part of your system, or extend it, but with the modern patented-design plastic systems you will have to stay with the same make for, though similar in style, they are not interchangeable.

Profiles
A diverse range of profiles is available, some based on classic profiles, others of modern form. See chart above, for typical sizes.

Plain half round

Ogee (OG)

Moulded OG

Box

MAINTENANCE

Cast iron, cast aluminium and asbestos cement guttering are all rigid and will support a ladder. But this is not really advisable, and it is much better to use a ladder stay (▷). Never prop a ladder against either plastic or roll-formed aluminium gutters.

Inspect and clean out the interiors of gutters regularly. Gutters concentrate the dirt, and sometimes sand washed down from the tiles by the rain. This can quickly build up if the flow of water is restricted by leaves or twigs. Birds' nests also can effectively block the guttering or downpipes.

The weight of standing water can distort plastic guttering, and if a blockage that is causing the gutter to overflow is left unchecked it can lead to problems with damp in the walls.

Removing debris
First block the gutter outlet with rag. With a shaped piece of plastic laminate scrape the silt into a heap, scoop it out of the gutter with a garden trowel and deposit it in a bucket hung from the ladder. Sweep the gutter clean with a stiff hand brush.

Remove the rag and flush the gutter down with a bucket of water.

Fit a wire or plastic 'balloon' to prevent debris falling into the downpipe or birds nesting on top of it.

Snow and ice
Plastic guttering can be badly distorted and even broken by snow and ice building up in it. Dislodge the build-up with a broom from an upstairs window if you can reach it safely. Otherwise climb a ladder to remove it.

If snow and ice become a regular seasonal problem you should screw a snow board made from 75 x 25mm (3 x 1in) planed softwood treated with a wood preservative and painted. Fix it to stand about 25mm (1in) above the eaves tiles, using 25 x 6mm (1 x ¼in) steel straps bent as required.

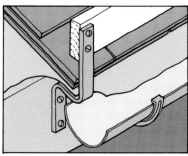
A snow board protects gutters or glazed roofs

GUTTERING MATERIALS

CAST IRON

The cast iron rainwater systems common on old houses are mostly of the OG type, fixed to the fascia board by mushroom-headed short screws through the back of the gutter above the water line and spaced about 600mm (2ft) apart.

Each 1.8m (6ft) standard length of the guttering has a socket end into which the plain spigot end of the next piece fits (1). Short bolts secure the joint and a bedding of putty forms a seal when they are tightened (2).

Cast iron is heavy and brittle, and installing or dismantling such a system needs two people. The iron can be cut with a hacksaw and drilled with standard twist drills in a power tool.

The guttering needs normal regular painting, and a bituminous paint applied inside makes for a longer life. If it is unprotected it will rust, usually along the back edge round the screws. Badly rusted guttering should be replaced, as it is likely to collapse.

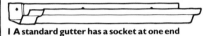
1 A standard gutter has a socket at one end

2 The joint is sealed with putty then bolted

CAST ALUMINIUM

Cast aluminium guttering comes in a wide range of profiles, including OG, moulded OG and half round. It is assembled in a similar way to cast iron guttering, with bolted joints, but a flexible mastic is used instead of putty to make the seals. The guttering may be fitted to the fascia with screws through the back, or with gutter brackets. All of the fixings should also be of aluminium.

Cast aluminium is about one third the weight of cast iron and can be left unpainted, but in some situations it can corrode, and if it is used as part of a cast iron system all the aluminium surfaces must be protected with zinc chromate or bituminous paint. It can be worked with ordinary metal-working tools.

ASBESTOS CEMENT

Asbestos cement guttering is of the socket-and-spigot type. The joints are secured with galvanised iron bolts and the seals are made with a mastic jointing compound. The guttering is produced in half round profiles in a range of several sizes. It is fixed to the house fascia board with gutter brackets, and these should not be spaced more than 900m (3ft) apart.

Asbestos cement has good weathering properties and does not need to be painted. It can be cut with a hacksaw and drilled normally (▷).

ROLLED SHEET ALUMINIUM

Rolled sheet aluminium guttering is a moulded lightweight OG system made from thin, prepainted flat sheet aluminium which is roll-formed to the gutter shape by a portable machine. This is done on site by the suppliers, and unbroken lengths are made to measure.

The end stops and angles are supplied as separate items, crimped to the ends of the gutter sections. Outlets are formed by punching holes in the bottom.

Simple metal fixing brackets are clipped to the front and back edges of the guttering and fixed to the fascia with drive screws. The system needs no maintenance but can be painted.

UPVC

Guttering of unplasticised PVC (uPVC) is now the most widely used, for new buildings and for replacing old systems. In many profiles and sizes, it is self-coloured in brown, black, grey and white. It needs no painting.

Most uPVC systems use clip-fastened joints with synthetic rubber gaskets to form the seals. Some use a solvent cement to weld the joints. Downpipes may be a push fit, and sealed with an '0' ring or solvent-welded. The guttering is usually supported by brackets, but some systems use screws in the back edge of the outlet and corner fittings for easier installation.

SEE ALSO
Details for: ▷

Ladder stay	78
Cutting asbestos	58

Cast iron OG gutter

Cast aluminium

RAFTER BRACKET　　FASCIA BRACKET
Gutter brackets

DRIVE SCREW
Sheet aluminium

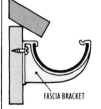
FASCIA BRACKET
Plastic guttering

FITTING NEW GUTTERING AND DOWNPIPE

When your old gutter system reaches the end of its useful life you should replace it. Try to do so with a system in the same style or, at least, one that *goes with the character of your house. If you plan to install the guttering yourself a plastic system is probably the best choice, being easy to handle.*

Installing the guttering

Measure round the base of the house to find the total length of gutter needed, and note the number and type of fittings to be ordered.

Use a plumb line to mark the position of the gutter outlet – directly over the existing drain – on the fascia board **(1)**. Screw the outlet or its support bracket – depending on the system – to the fascia no more than 50mm (2in) below the tile level **(2)**.

Fix a gutter bracket at the opposite end of the run, close to the top of the fascia board. This is to give a fall of at least 25mm (1in) in 15m (50ft) **(3)**. Run a taut string between the bracket and outlet and fix the rest of the brackets along the slope of the string, spaced no more than 1m (3ft 3in) apart **(4)**. There should be a bracket at every joint unless the system uses screw-fixed outlets and angles **(5)**.

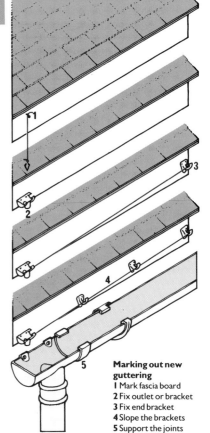

Marking out new guttering
1 Mark fascia board
2 Fix outlet or bracket
3 Fix end bracket
4 Slope the brackets
5 Support the joints

Fitting the gutter

Tuck the back edge of a length of gutter under the roofing felt and into the rear lips of the brackets, then clip its front edge into all of the brackets **(6)**.

Fit the second length in the same way, with its spigot end pushed into the socket of the first length. Some pressure will be needed to compress the rubber seal. Leave a gap of about 6mm (¼in) between the end of the spigot and the shoulder of the socket to allow for expansion.

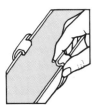

6 Clip the guttering into the brackets

7 Use an adaptor to join different systems

● **Cutting asbestos**
To cut asbestos, wet it to damp-down the dust and wear a face mask. Place plastic sheet under the work to catch falling dust. Carefully fold the sheet and pick up any spilt dust with a damp sponge. Seal it in a plastic bag. Your local authority will advise you on disposal.

Cutting the gutter

Cut the gutter squarely with a hacksaw. You can snap a clip or bracket over it first to give rigidity and guide the cut. Smooth the rough edges with a file. In some systems notches for the clips are made in the gutter's front and back edges; this can be done with a file.

Connecting to existing guttering

To renew guttering on a terrace house may mean joining your system to your neighbours'. There are left- and right-hand adaptors for this.

Remove your old guttering to the nearest joint between the houses, bolt the adaptor on, sealing the joint with mastic, and fix the new plastic gutter with the clip provided. **(7)**.

Fitting the downpipe

Work downward from the gutter outlet. If the eaves overhang you will need to fit an offset.

Fit a clip to the top of a length of downpipe. Hold it against the wall and measure the distance from its centre to a plumb line dropped through the outlet's centre **(8)**. You may find an offset to fit but will most likely have to make it up with offset bends and pipe. Use a solvent cement and assemble it on a table so that the bends lie in the same plane.

Fit the offset to the outlet spigot and the pipe to the offset. Adjust the pipe so that the clip's back plate falls on a mortar joint **(9)**. Trim the offset spigot if necessary.

Mark the fixings, drill and plug the wall and fix the pipe and clip with plated round-head screws.

Mark and fix the lower lengths of pipe in the same way, with a 6mm (¼in) expansion gap between each pipe and the socket shoulder. Fit extra clips at the centre of any pipe longer than 2m (6ft 6in).

Cut the bottom pipe to length and fit a shoe in the same way **(10)**. If the pipe is jointed into a gulley trap **(11)** or a drainsocket you may have to work from the bottom.

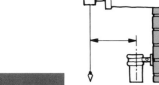 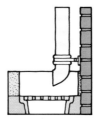 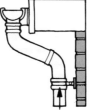 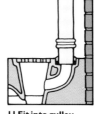

8 Drop a plumb line   9 Fit an offset   10 Finish with a shoe   11 Fit into gulley

REPAIRING GUTTERS AND DOWNPIPES

It is always better to replace a damaged part of a gutter system than to repair it, but a neat repair that stays watertight can be regarded as permanent.

Mending a crack

Chipped or cracked guttering and downpipes of cast iron or asbestos cement can be repaired with a 'cement bandage'.

Clean the surface, cut the bandage to length and soak it in water for 15 to 20 seconds. Wearing rubber gloves position the bandage over the crack, smooth it in place and leave it to set hard. If possible cover it with polythene for 72 hours to delay the drying time.

Using epoxy putty

You can build up the chipped edge of a cast iron gutter with epoxy putty. Tape a waxed or polythene-lined piece of card across the outside of the broken edge and bent to follow the contour of the gutter. Mix the two-part putty, fill the gap with it and remove the cardboard 'former' when the putty sets.

INSULATION

INSULATION
ROOFS

About a quarter of the heat lost from an average house goes through the roof, so preventing this should be one of your priorities when it comes to insulating your home. Providing you're able to gain access to your loft floor, reducing substantial heat loss is just a matter of laying the insulation material between the joists: it's cheap, quick and effective. If you want to use the attic, insulating the sloping surface of the roof is a quite straightforward alternative.

Treating a flat roof

A flat roof – on an extension for instance – may also need insulating, but the only really practical solution for most householders is to apply a layer of insulation to the ceiling surface. It is not a particularly difficult task, providing the area is not too large, but you will have to re-locate lighting and take into consideration features such as cupboards or windows that extend to the ceiling. Fixing ceiling tiles is an alternative but their thermal-insulation value is minimal.

Preparing the loft

On inspection, you may find that the roof space has existing but inadequate insulation. At one time even 25mm (1in) of insulation was considered to be acceptable. It is worth installing extra insulation to bring it up to the recommended thickness of 100mm (4in). Check roof timbers for woodworm or signs of rot so that they can be treated first (▷). Make sure that the electrical wiring is sound; lift it clear so that you can lay insulation beneath it.

Plaster or plasterboard ceiling surfaces will not support your weight, so lay a plank or two, or a panel of chipboard, across the joists so that you can move about safely: don't allow it to overlap the joists; if you step on the edge it will tip over.

If there is no permanent lighting in the loft, rig up an inspection lamp on an extension lead, so you can move it wherever it is needed, or hang it high up for best overall light.

Most attics are very dusty, so wear old clothes and a gauze facemask. You may wish to wear protective gloves, particularly if you are handling glass-fibre batts or blanket insulation, which can irritate sensitive skin.

TYPES OF LOFT INSULATION

There's a wide range of different insulation materials available. Check out the recommended types with your local authority before you apply for a grant.

BLANKET INSULATION

Glass-fibre and mineral- or rock-fibre blanket insulation is commonly sold as rolls made to fit snugly between the joists. The same material cut to shorter lengths is also sold as 'batts'. A minimum thickness of 100mm (4in) is recommended for loft insulation. Blanket insulation may be unbacked, paper-backed to improve its tear-resistance, or it may have a foil backing as a vapour barrier (See below).

The unbacked type is normally used for laying on the loft floor. Blankets are usually 75mm (3in) or 100mm (4in) thick and typically 400mm (1ft 4in) wide – suitable for the normal joist spacing. For wider than usual joist spacing, choose the 600mm (2ft) width (cut it in half for narrow joist spaces with a panel saw before you unwrap it). Rolls are typically 6 to 8m (20 to 25ft) long.

If you want to fit blanket insulation to the sloping part of the roof, buy it with a lip of backing along each side for stapling to the rafters.

Both glass and mineral fibre are non-flammable and proofed against damp, rot and vermin.

LOOSE-FILL INSULATION

Loose-fill insulation in pellet or granular form is poured between the joists on the loft floor to a minimum depth of 100mm (4in), although a depth of 130mm (5¼in) is recommended for the same value of insulation as 100mm (4in) thick blanket – but this could rise above some joists.

Exfoliated vermiculite, made from a mineral called mica, is the most common form of loose-fill insulation but others such as mineral wool, polystyrene or cork granules may be available. Loose-fill is sold in bags containing 110 litres (4cu ft) – sufficient to cover 1.1sq m (12sq ft) at 100mm (4in) deep.

It's inadvisable to use loose-fill in a draughty, exposed loft: high winds can cause it to blow about. However, it's convenient to use if the joists are irregularly spaced.

BLOWN FIBRE INSULATION

Professional contractors can provide inter-joist loft insulation by blowing glass, mineral or cellulose fibres through a large hose. A minimum, even depth of 100mm (4in) is required. Blown fibre insulation may be unsuitable for a house in a windy location but seek the advice of the contractor.

RIGID AND SEMI-RIGID SHEET INSULATION

Sheet insulation is principally for fixing between the rafters. You can choose semi-rigid batts of glass or mineral fibre, or fibre insulation board. A minimum thickness of 25mm (1in) is required when covered with plasterboard, but install thicker insulation where possible, allowing for sufficient ventilation between it and the roof tiles or slates to avoid condensation (▷).

SEE ALSO

Details for: ▷	
Woodworm	16
Dry and wet rot	15
Condensation	10
Roof ventilation	71–72

ESTIMATING FOR BLANKET INSULATION
400mm wide rolls

Approx. loft area

Square metres	Square feet	No. of rolls
29	314	13
31.5	339	14
34	363	15
36	387	16
38	411	17
40.5	435	18
43	460	19
45	487	20
56	605	25
67.5	726	30
79	847	35
90	968	40

Allows for average joist widths of 50mm (2in)

VAPOUR BARRIERS

Installing insulation has the effect of making the areas of the house outside that layer of insulation colder than before, so increasing the risk of condensation (▷) either on or within the structure itself. In time this could result in decreased value of the insulation and may promote a serious outbreak of dry rot in household timbers (▷).

To prevent this happening, it's necessary to provide adequate ventilation for those areas outside the insulation or to install a vapour barrier on the inner, or warm, side of the insulation to prevent moisture-laden air passing through. This is usually a plastic sheet or layer of metal foil, which is sometimes supplied along with the insulation. It is essential that a vapour barrier is continuous and undamaged or its effect is greatly reduced.

● **Ventilating the loft**
Laying insulation between the joists increases the risk of condensation in an unheated roof space above, but, provided there are gaps at the eaves, there will be enough air circulating to keep the loft dry (▷).

59

INSULATING THE LOFT

Laying blanket insulation

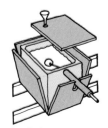

Insulating a tank
Cut slabs of insulant to fit sides: peg corners. Fit lid plus a funnel to collect drips from expansion pipe (See box right)

Seal gaps around pipes, vents or wiring entering the loft with flexible mastic. Remove the blanket wrapping in the loft (it's compressed for storage and transportation but swells to its true thickness when released) and begin by placing one end of a roll into the eaves. Make sure you don't cover the ventilation gap – trim the end of the blanket to a wedge-shape so it does not obstruct the airflow, or fit eaves vents.

Unroll the blanket between the joists, pressing it down to form a snug fit, but don't compress it. If the roll is slightly wider than the joist spacing, allow it to curl up against the timbers on each side.

Continue at the opposite side of the loft with another roll: cut it to butt up against the end of the first one, using a large kitchen knife or long-bladed pair of scissors. Continue across the loft until all the spaces are filled. Cut the insulation to fit odd spaces.

Do not cover the casing of any light fittings which protrude into the loft space. Avoid covering electrical cables, as there's a risk it may cause overheating. Instead, lay the cables on top of the blanket, or clip them to the sides of the joists above it.

Do not insulate the area directly below a cold water tank, so that heat rising from the room below will help to prevent freezing. Cut a piece of insulation to fit the hatch cover and attach it with PVA adhesive or hold it down with cloth tapes and drawing pins. Fit foam draught excluder around the edge of the hatch.

Laying loose-fill insulation

Take similar precautions against condensation to those described for blanket insulation. To prevent blocking the eaves, wedge strips of plywood or thick cardboard between the joists. Pour insulation between the joists and distribute it roughly with a broom. Level it with a spreader cut from hardboard to fit between the joists: notch it to fit over the joists so the central piece levels the granules accurately.

To insulate the entrance hatch, screw battens around the outer edge of the cover, fill with granules and pin on a hardboard lid to contain them.

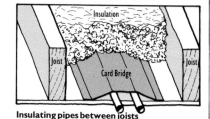

Insulating pipes between joists

INSULATING TANKS AND PIPES

Insulating tanks
Insulate the cold water storage tanks in the loft, including the central heating expansion tank. Buy a ready-made kit to fit the tank or make one from suitable insulation.

You can construct a hardboard liner to contain loose-fill insulation, but it's easier to surround the tank with sheets of 25mm (1 in) thick expanded-polystyrene or mineral-fibre slabs. Cut slots for the pipework and make a lid to cover the tank. Fit a plastic funnel in the lid to catch drips from the expansion pipe (◁). Join the corners with sharpened wooden dowel pegs or tie with string, wire or tape.

Alternatively, lag the tank with blanket insulation but cut a rigid lid to fit and cover it with insulation; wrap with polythene to prevent fibres contaminating the water.

Insulating pipes
If cold water pipes run between the joists, lay the blanket insulation over them to ensure they will not freeze. If this is not practical, insulate all pipe runs separately (◁).

Lay a thin card bridge over cold water pipework running between the joists before pouring loose insulation, so that the pipes benefit from warmth rising from the room below. If the joists are shallow, cover the pipes with foam sleeves before pouring the insulation.

Left
Seal gaps around pipes and vents (1). Place end of roll against eaves and trim ends (2) or fit eaves vents (3). Press rolls between joists (4) Insulate tank and pipes (5).

Right
Seal gaps to prevent condensation (1). Stop insulant blocking ventilation with strips of plywood (2) or eaves vents (3). Cover cold water pipes with a cardboard bridge (4) then use a spreader to level the insulant (5). Insulate and draughtproof the hatch cover (6).

Laying blanket insulation in the loft

Spreading loose-fill insulant in the loft

INSULATING A SLOPING ROOF

Insulating between the rafters

If the attic is in use, you will need to insulate the sloping part of the roof in order to heat the living space. Repair tiles or slates first, as not only will leaks soak the insulation but also it will be difficult to spot them after insulating.

Condensation is a serious problem when you install insulation between the rafters, as the undersides of the tiles will become very cold. You must provide a 50mm (2in) gap between the tiles and the insulation to promote sufficient ventilation to keep the space dry, which in turn determines the maximum thickness of insulation you can install. The ridge and eaves must be ventilated (▷) and you should include a vapour barrier on the warm side of the insulation, either by fitting foil-backed blanket or by stapling sheets of polythene to the lower edges of the rafters to cover unbacked insulation.

Whatever insulation you decide on, you can cover the rafters with sheets of plasterboard as a final decorative layer. The sizes of the panels will be dictated by the largest boards you can pass through the hatchway. Use plasterboard nails or screws to hold the panels against the rafters, staggering the joints (▷). Alternatively, provide insulation and surface finish together by fitting insulated (thermal) plasterboard to the underside of the rafters (▷).

Fixing blanket insulant

Unfold the side flanges from a roll of foil-backed blanket and staple them to the underside of the rafters. When fitting adjacent rolls, overlap the edges of the vapour barrier to provide a continuous layer.

Attaching sheet insulant

The simplest method is to cut the sheet insulation accurately so that it will be a wedge-fit between the rafters. If necessary, screw battens to the sides of the rafters to which you can fix the insulating sheets. Treat the battens beforehand with preservative. Staple a polythene sheet vapour barrier over the rafters. Double-fold the joints over a rafter and staple in place.

INSULATING AN ATTIC ROOM

If an attic room was built as part of the original dwelling, it will be virtually impossible to insulate the pitch of the roof unless you are prepared to hack off the old plaster and proceed as left. It may be simpler to insulate from the inside as for a flat roof (▷) but your headroom may be seriously hampered.

Insulate the short vertical wall of the attic from inside the crawlspace, making sure the vapour barrier faces the inner, warm side of the partition. Insulate between the joists of the crawlspace at the same time.

Fit blankets with vapour barrier facing the room

SEE ALSO

Details for: ▷	
Insulating a flat roof	62
Thermal board	63
Ventilating ridge/eaves	71
Preservatives	16

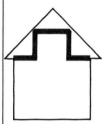

Insulating a room in the attic
Surround the room itself with insulation but leave the floor uninsulated so that the room benefits from rising heat generated by the space below.

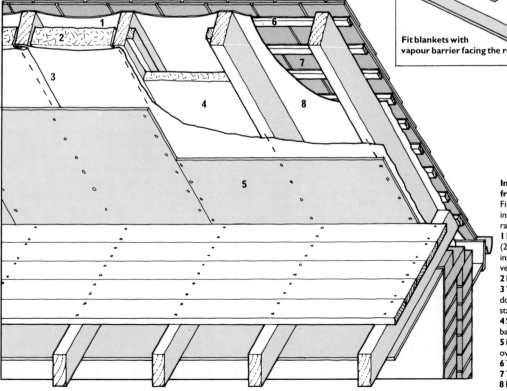

Insulating an attic from the inside
Fit blanket or sheet insulant between the rafters.
1 Minimum of 50mm (2in) between insulation and tiles for ventilation
2 Blanket or batts
3 Vapour barrier with double-folded joints stapled to rafters
4 Sheet insulant fixed to battens
5 Plasterboard nailed over vapour barrier
6 Tile battens
7 Tiles or slates
8 Roof felt

INSULATING A FLAT ROOF

INSULATING WALLS

Treatment from above

A flat roof can be insulated from the outside by laying rigid insulating board on the original deck and weighting it down with paving slabs or a layer of pebbles. This is a job for a professional only, but insist that the contractor checks that the roof can support the additional weight and that the insulation is waterproofed and sealed all round to prevent penetrating damp.

Insulating a flat roof from outside
A professional company can insulate the roof from above.
1 Roof deck
2 Waterproof covering
3 Insulation
4 Paving slabs to hold insulant in place

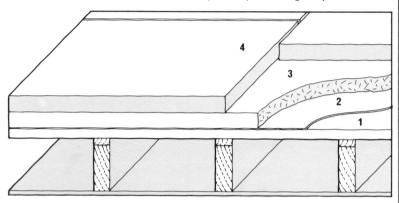

Treatment from below

The only other option for insulating a flat roof is to treat the ceiling below; you can do this so long as you have sufficient headroom. You must guard against condensation in the restricted space within the roof structure, particularly as it is often insufficiently ventilated – if at all. You should include a polythene vapour barrier on the warm side of the ceiling, below the insulation.

Fire-retardant expanded polystyrene 25mm (1in) thick is normally recommended for its lightness, with softwood battens screwed to the joists between the panels of insulation every 400mm (1ft 4in) across the ceiling. Fit the first batten against the wall at right-angles to the joists, and one at

each end of the room. Butt the polystyrene against the first batten, gluing it to the ceiling with special polystyrene adhesive. Coat the entire back of the material.

Continue with alternate battens and panels until you reach the opposite side of the room, finishing with a batten against the wall. Staple a polythene vapour barrier across the framework, double-folding the joins, over the nearest batten. Fix plasterboard panels to the battens with galvanized plasterboard nails. Stagger the joins between the panels then fill the joins and finish ready for decorating as required. A double coat of oil paint is a vapour barrier itself.

Insulating the ceiling
Insulate a flat roof by fixing insulant to the ceiling.
1 Existing ceiling
2 Softwood battens screwed to joists
3 Insulation glued to existing ceiling
4 Polythene vapour barrier stapled to battens
5 Plasterboard nailed to battens

How you insulate the walls of your home will be determined by several factors. Firstly, the type of construction. If the house was built since 1920 and certainly after 1950, it will probably have cavity walls – two skins of brick, or one of brick and one of concrete block, with a gap between through which air circulates to reduce the likelihood of water penetration. Although heat loss is slightly slower through a cavity wall than one of solid brick, it is not sufficiently insulating to substantially reduce the cost of home heating. However, filling the cavity with insulation prevents circulation, trapping the air in millions of tiny air pockets within the material: it's a process that can cut heat loss through the wall by about 65 per cent.

Solid walls require different treatment: basically, you can either insulate the external face of the walls or line the inner surface – and either of these methods can involve considerable disruption to the joinery, electrical and plumbing supplies.

Advantages and disadvantages

With cavity filling, every exterior wall must be treated simultaneously, so it is most cost-effective for homes which are heated throughout for long periods and with a properly controlled system. Heating without controls will simply increase the temperature inside instead of saving on fuel bills. This type of insulation is impractical for flats or apartments unless the whole building is insulated at the same time.

Solid masonry walls must be insulated in some other way. It is possible to hire a contractor to clad the exterior of the house with insulation (◁) but it is expensive and alters the appearance of some buildings. The comments concerning the manner of heating and effective controls to make cavity filling worthwhile apply equally to exterior wall insulation.

Another method is to line the inner surface of either type of wall construction with insulation. It may involve a great deal of effort depending on the amount of alteration required to joinery, electrical fittings, and so on, but it provides the opportunity for selective insulation, concentrating on those rooms which would benefit most, and it is the only form of wall insulation which can be carried out by the householder.

INSULATING A CAVITY WALL

When constructing a new house, a builder will include a layer of insulation between the two masonry leaves of exterior walls: a simple measure, which greatly increases the thermal insulation of the building. To insulate an existing wall is a different matter. It requires a skilled and experienced contractor to introduce an insulant through holes cut in the outer brick leaf to fill the cavity in such a way that you achieve the savings in heat loss without the side effects of damp penetrating to the inner leaf.

Hire a contractor who is approved by the Agrément Board, registered with the British Standards Institution, or who is a member of the National Cavity Insulation Association.

You should expect the contractor to carry out a thorough survey of the building to make sure the walls are structurally fit for filling, with no evidence of frost damage or failed pointing. An approved company will also make the necessary application to the local authority before commencing installation to comply with the Building Regulations. This is particularly important if you live in an area of the country where your house is exposed to severe driving rain for prolonged periods. Not all cavity fillings are suitable for such extreme weather conditions.

Do not hire a firm which does not provide a long-term guarantee, which is transferable with home ownership. It should state that the insulant will be effective for the period of the guarantee and that it is rot- and vermin-proof. Be especially careful to check that the contractor's guarantee states that damp resulting from faulty material or installation will be cured free of charge.

More than likely you will be offered one of three insulants: urea formaldehyde foam is the cheapest and most widely used insulant, despite its reputation for releasing unpleasant odours as the foam cures. In reality, it happens in very few instances and is due to a poor survey, which did not detect that the walls were not in a fit condition for filling.

Mineral or glass fibre treated with water repellent are the next most popular cavity insulants. Both materials are completely inert and if properly installed, will form a stable insulation that will neither settle nor shrink once installed in the cavity.

Expanded-polystyrene beads form the third most commonly used insulant. Some of them are lightly coated with adhesive at the moment of injection so that the fill does not settle over a period of time. If polystyrene is treated and properly installed it does not affect the fire resistance of a masonry wall.

Whatever process you choose, it should take no more than two to three days to complete. All the work is carried on outside the house, where holes are drilled at regular intervals in the brickwork (1). The insulant is injected or blown through a hose (2) and finally the holes are plugged. If the work is done properly, the holes should be virtually invisible, except perhaps by close inspection.

1 Drilling holes in the outer leaf
A professional contractor will begin by drilling large diameter holes through the outer skin of brickwork into the cavity.

2 Introducing the insulant
A hose is inserted into each hole and the insulant is injected or blown under pressure into the cavity, filling it from the base. When the job is complete, the holes are plugged with colour-matched mortar.

DRY-LINING FROM THE INSIDE

If you are planning to dry-line an external wall with some form of panelling (▷), take the opportunity to include blanket or sheet insulation between the wall battens or furring strips (▷). Fix a polythene vapour barrier over the insulation by stapling it to the furring strips before nailing the panelling in place.

Alternatively, use a metallized plastic-backed plasterboard, which requires no additional vapour barrier.

Any form of panelling can be applied over mineral- or glass-fibre blanket but plasterboard should be used to cover expanded-polystyrene insulant.

A simpler method is to glue insulated (thermal) plasterboard directly to a sound plaster surface. This type of wall lining is standard plasterboard, backed with a layer of expanded-polystyrene or rigid-polyurethane foam. An integral vapour barrier is incorporated in both boards (see below).

Using a notched applicator, apply a 200mm (8in) wide band of the manufacturer's adhesive to the wall to coincide with the vertical edges of the panel and its centre line (1). Spread horizontal bands top and bottom. Press the panel against the adhesive and tamp it down with a heavy straightedge then secure it to the wall with nine nailable plugs (2) in three rows. Position the plugs 50mm (2in) from all edges of the panel and one in the centre. Cut a panel to fit into a corner with a fine-toothed saw. Allow for a 3mm (⅛in) gap at a cut edge for filling after the panel is fixed. Tape and fill all joints.

Rebuild the door and window mouldings to cover the cut edge of the dry-lining panels. Bed the skirting board in a bed of mastic sealant applied to the floor and plasterboard lining, then fix the skirting through the thickness of the panel to the wall behind.

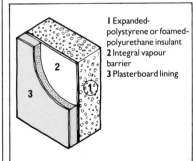

1 Expanded-polystyrene or foamed-polyurethane insulant
2 Integral vapour barrier
3 Plasterboard lining

The structure of insulated plasterboard

SEE ALSO

Details for: ▷
| Furring strips | 78 |
| Panelling | 78 |

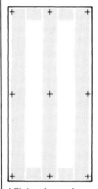

1 Fixing thermal plasterboard
Spread adhesive along the bands shown in the diagram above. The crosses indicate the centres of nailable fixing plugs

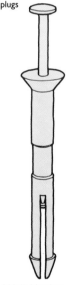

2 Nailable plug
Used to fix insulated plasterboard to the wall. Push the plug into a hole drilled through the board, then drive in the nail to expand the plug and grip the masonry.

LAGGING PIPES, CYLINDER AND RADIATORS

Insulating a hot water cylinder

Many people think that an uninsulated cylinder is providing a useful source of heat in an airing cupboard, but in fact it squanders a surprising amount of energy. Even a lagged cylinder should provide ample background heat in an enclosed cupboard – but if not, an uninsulated pipe will.

Proprietary cylinder jackets are made from segments of 80 to 100mm (3¼ to 4in) thick mineral-fibre insulation material wrapped in plastic. Measure the approximate height and circumference of the cylinder to choose the right size. If necessary, buy a larger jacket rather than one that is too small. Make sure you buy a good-quality jacket by checking that it is marked with the British Standard kite mark (BS 5615).

If you should ever have to replace the cylinder, consider buying a pre-insulated version, of which there are various types on the market

Thread the tapered ends of the jacket segments onto a length of string and tie it round the pipe at the top of the cylinder. Distribute the segments evenly around the cylinder and wrap the straps or string provided around it to hold the insulation in place.

Spread out the segments to make sure the edges are butted together and tuck the insulation around the pipes and the cylinder thermostat.

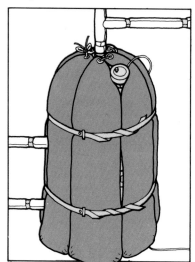

Lagging a hot water cylinder
Fit a jacket snugly around the cylinder and wrap insulating foam tubes (See above right) around the pipework, especially the vent pipe directly above the cylinder.

Lagging pipe runs

You should insulate hot water pipes where their radiant heat is not contributing to the warmth of your home, and also cold water pipe runs in unheated areas of the house, where they could freeze. You can wrap pipework in one of several lagging bandages, some of which are self-adhesive, but it is more convenient to use foamed plastic tubes designed for the purpose, especially along pipes running close and clipped to a wall, which would be awkward to wrap.

Plastic tubes are made to fit pipes of different diameters and the tube walls vary in thickness from 10mm to 20mm (½in to ¾in). More expensive tubing incorporates a metallic foil backing to reflect some of the heat back into hot water pipes.

Most tubes are pre-slit along their length so that they can be sprung over the pipe (**1**). Butt successive lengths of tube end-to-end and seal the joints with PVC adhesive tape.

At a bend, cut small segments out of the split edge so it bends without crimping. Fit it around the pipe (**2**) and seal the closed joints with tape. If pipe is joined with an elbow fitting, mitre the ends of the two lengths of tube, butt them together (**3**) and seal with tape.

Cut lengths of tube to fit completely around a tee-joint, linking them with a wedge-shaped butt joint (**4**) and seal with tape as before.

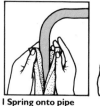

I Spring onto pipe   **2 Cut to fit bend**   **3 Mitre over elbows**   **4 Butt at tee-joint**

Reflecting heat from a radiator

Up to 25 per cent of the radiant heat from a radiator on an outside wall is lost to the wall behind. Reclaim perhaps half this wasted heat by installing a foil-faced, expanded polystyrene lining behind the radiator to reflect it back into the room.

The material is available as rolls, sheets or tiles to fit any size and shape of radiator. It is easier to stick the foil on the wall when the radiator is removed for decorating but it is not an essential requirement.

Turn off the radiator and measure it, including the position of the brackets. Use a sharp trimming knife or scissors to cut the foil to size so that it is slightly smaller than the radiator all round. Cut narrow slots to fit over the fixing brackets (**1**).

Apply heavy-duty fungicidal wallpaper paste to the back of the sheet and slide it behind the radiator (**2**). Rub it down with a radiator roller or smooth it against the wall with a wooden batten. Allow enough time for the adhesive to dry before turning the radiator on again.

I Cut slots to align with wall brackets

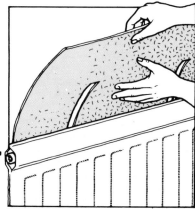

2 Slide lining behind radiator and press to wall

A double-glazed window consists of two sheets of glass separated by an air gap. The air gap provides an insulating layer, which reduces heat loss and sound transmission. Condensation is also reduced because the inner layer of glass remains relatively warmer than that on the outside. Factory-sealed units and secondary glazing are the two methods in common use for domestic double glazing. Both will provide good thermal insulation. Sealed units are unobtrusive, but secondary glazing can offer improved sound insulation. But which do you choose to suit your house and your lifestyle?

What size air gap?

For heat insulation a 20mm (¾in) gap provides the optimum level of efficiency. Below 12mm (½in) the air can conduct a proportion of the heat across the gap. Above 20mm (¾in), there is no appreciable extra gain in thermal insulation and air currents can occur, which transmit heat to the outside layer of glass. A larger gap of 100 to 200mm (4 to 8in) is more effective for sound insulation. A combination of a sealed unit plus secondary glazing provides the ideal solution, and is known as triple glazing.

Double glazing will help to cut fuel bills but its immediate benefit will be felt by the elimination of draughts. The cold spots associated with a larger window, particularly noticeable when sitting relatively still, will also be reduced. In terms of saving energy, the heat lost through windows is relatively small – around 10 to 12 per cent – compared to the whole house. However, the installation of double glazing can halve this amount.

Double glazing will improve security against forced entry, particularly if sealed units or toughened glass have been used. However, ensure some accessible part of the window is openable to provide emergency escape in case of fire.

SEE ALSO

Details for: ▷	
Secondary double glazing	66, 68
Plastic glazing	67

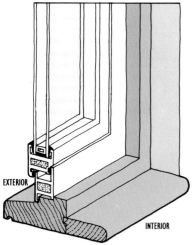

Factory-sealed unit
A complete frame system installed by a contractor.

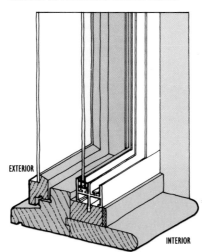

Secondary window system
Fitted in addition to the normal glazed window.

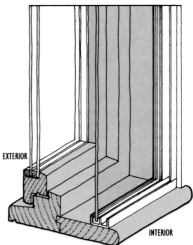

Triple glazing
A combination of secondary and sealed units.

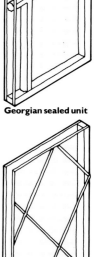

Georgian sealed unit

Leaded light unit

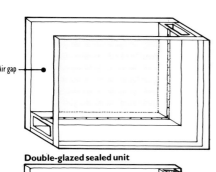

Double-glazed sealed unit

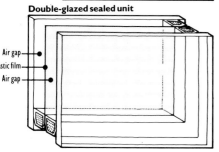

Heat-retentive sealed unit

Double-glazed sealed units

Double-glazed sealed units are manufactured from two panes of glass separated by a spacer and hermetically sealed all round. The cavity between the glass may be 5, 9, 12 or 20mm (¼, ⅜, ½ or ¾in) wide. The gap may contain dehydrated air to eliminate condensation between the glass, or inert gases which also improve thermal and accoustic insulation.

The thickness and type of glass is determined by the size of the unit. Clear float glass or toughened glass is common. When obscured glazing is required for privacy, patterned glass is used. Special heat-retentive sealed units are also supplied by some double glazing companies, incorporating special glass or a plastic film embodied within the unit.

A leaded light and a Georgian version of the sealed unit are also produced. The former is made by bonding strips of lead to the outer pane of the glass, the latter by placing a moulded framework of glazing bars in the cavity. The improved security, lack of maintenance and ease of cleaning in some way make up for their lack of character compared with the original style of window.

Generally these special sealed units are produced and installed by suppliers of ready-made double-glazed replacement windows. Sealed units are available for self-fixing from some joinery suppliers or they can be made to order by specialists. Square-edged units are made for frames with a deep rebate, and stepped units for frames intended for single glazing

SECONDARY DOUBLE GLAZING

Secondary double glazing comprises a separate pane of glass or plastic sheet which is normally fitted to the inside of existing single-glazed windows. It is a popular method for double glazing windows, being relatively easy for home installation – and usually at a fraction of the cost of other systems.

How the glazing is fixed

Glazing can be fastened to the sash frames (**1**), the window frames (**2**), or across the window reveal (**3**). The choice depends on the ease of fixing, the type of glazing and personal requirements for ventilation.

Glazing fixed to the sash will cut down heat loss through the glass and provide accessible ventilation, but it will not stop draughts. That fixed to the window frame will reduce heat loss and stop draughts at the same time. Glazing fixed across the reveal will also offer improved sound insulation as the air gap can be wider. Any system should be readily demountable or preferably openable to provide a change of air in a room without some other form of ventilation.

Rigid glazing of plastic or glass can be fitted to the exterior of the window opening if secondary glazing would spoil the appearance of the interior. In this case, windows which are set in a deep reveal, such as the vertically sliding sash type, are the most suitable (**4**).

Glazing with renewable film

Effective double glazing can be achieved using double-sided adhesive tape to stretch a thin, flexible sheet of plastic across the window frame. It can be removed at the end of the cold season without harming the paintwork.

Clean the window frame (**1**) then cut the plastic sheet roughly to size, allowing an overlap all round. Apply double-sided tape to the frame edges (**2**) and peel off its backing paper.

Attach the film to the top rail (**3**), then tension it onto the tape on the sides and bottom of the frame (**4**). Apply light pressure only until the film is positioned then rub it down onto the tape all round.

Use a hair dryer set to a high temperature to remove all creases and wrinkles in the film (**5**). Starting at an upper corner, move the dryer slowly across the film, holding it about 6mm (¼in) from the surface. When the film is tensioned, cut off the excess plastic (**6**).

GLAZING POSITIONS

Secondary double glazing is particularly suitable for DIY installation, partly because it is so versatile. It is possible to fit a system to almost any style or shape of window.

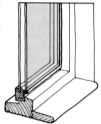

1 Sash fixed
Glazing fixed to the opening window frame

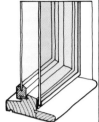

2 Frame fixed
Glazing fixed to the structural frame

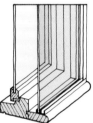

3 Reveal fixed
Glazing fixed to the reveal and interior window sill

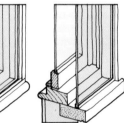

4 Exterior fitted
Glazing fixed to the reveal and exterior window sill

1 Wipe woodwork to remove dust and grease

2 Apply double-sided tape to the fixed frame

3 Stretch the film across the top of the frame

4 Pull the film tight and fix to sides and bottom

5 Use a hair dryer to shrink the film

6 Trim the waste with a sharp knife

Demountable systems

A simple method for interior secondary glazing uses clear plastic film or sheet. These lightweight materials are secured by self-adhesive strips or rigid moulded sections, which form a seal. Most strip fastenings use magnetism or some form of retentive tape, which allows the secondary glazing to be removed for cleaning or ventilation. The strips and tapes usually have a flexible foam backing, which takes up slight irregularities in the woodwork. They are intended to remain in place throughout the winter and be removed for storage during the summer months.

Fitting a demountable system

Clean the windows and the surfaces of the window frame. Cut the plastic sheet to size. Place the glazing on the window frame and mark around it (1). Working with the plastic on a flat table, peel back the protective paper from one end of the self-adhesive strip. Tack it to the surface of the plastic, flush with one edge. Cut it to length and repeat on the other edges. Cut the mating parts of the strips and apply them to the window frame following the guidelines. Press the glazing into place (2).

When dealing with rigid moulded sections, cut the pieces to length with mitred corners. Fit the sections around the glazing, peel off the protective backing and press the complete unit against the frame (3).

For economy and safety, plastic sheet materials can be used in place of glass to provide lightweight double glazing. They are available in clear thin flexible films or clear, textured and coloured rigid sheets.

Unlike glass, plastic glazing has a high impact-resistance and will not splinter when broken. Depending on thickness, plastic can be cut with scissors, drilled, sawn, planed and filed.

The clarity of new plastics is as good as glass but they will scratch. They are also liable to degrade with age and are prone to static. Plastic sheet should be washed with a liquid soap solution. Slight abrasions can be rubbed out with metal polish.

Film and semi-rigid plastics are sold by the metre or in rolls. Rigid sheets are available in a range of standard sizes or can be cut to order. Rigid plastic is covered with a protective film of paper or thin plastic on both faces which is peeled off only after cutting and shaping to keep the surface scratch-free.

POLYESTER FILM

A plastic film for inexpensive double glazing. It can be trimmed with scissors or a knife and fixed with self-adhesive tape or strip fasteners.

It is a tough, virtually tearproof film, which is very clear – ideal, in fact, for glazing living rooms. It is sold in 5, 10 and 25 metre (32, 64 and 160ft) rolls, 1143mm (3ft 9in) and 1300mm (4ft 3in) wide.

POLYSTYRENE

A relatively inexpensive clear or textured rigid plastic. Clear polystyrene does not have the clarity of glass and will degrade in strong sunlight. It should not be used for south-facing windows or where a distortion-free view is required. Depending on climatic conditions, the life of polystyrene is reckoned to be between three and five years. Its working life can be extended if the glazing is removed for storage in summer. It is available in thicknesses of 1.5mm ($\frac{1}{16}$in), 2.5mm ($\frac{3}{32}$in) and 4mm ($\frac{5}{32}$in) and sheet sizes up to 1220mm × 2440mm (4ft × 8ft).

ACRYLIC

A good-quality rigid plastic with the clarity of glass. It costs about the same as glass and about half as much again as polystyrene. Its working life is considered to be at least 10 years. Acrylic is also available in a wide range of translucent and opaque colours. The common thicknesses available for clear glazing are 1.5mm ($\frac{1}{16}$in), 2.5mm ($\frac{3}{32}$in) and 4mm ($\frac{5}{32}$in) and it is available in sheet sizes up to 1220mm × 2440mm (4ft × 8ft).

POLYCARBONATE

A relatively new plastic glazing material, which is virtually unbreakable. It provides a lightweight vandal-proof glazing with a high level of clarity. The standard grade costs about twice the price of acrylic. It is made in clear, tinted, opal and opaque grades, some with textured surfaces.

Thicknesses suitable for domestic glazing are 2mm ($\frac{1}{16}$in), 3mm ($\frac{1}{8}$in) and 4mm ($\frac{5}{32}$in), although greater thicknesses are made, some in grades which are even bullet-proof. Sheet size can be up to 2050mm × 3000mm (7ft × 10ft).

PVC GLAZING

PVC is available as a flexible film or as a semi-rigid sheet. The film provides inexpensive glazing where a high degree of clarity is not required such as in a bedroom. PVC is ultraviolet-stabilized so is suitable for outside or inside use.

Consequently, it is very suitable for glazing conservatories (and carport roofs). Rigid PVC is 3mm ($\frac{1}{8}$in) thick in sheet sizes up to 1220mm × 2440mm (4ft × 8ft).

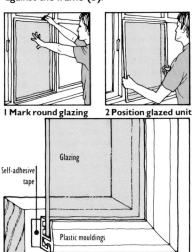

1 Mark round glazing   2 Position glazed unit

Glazing

Self-adhesive tape

Plastic mouldings

3 Rigid plastic mouldings support the glazing

Hinged or sliding secondary glazing systems are available in kit form for home assembly. Sliding systems are also made and installed by glazing companies – and both sliding and hinged types are intended to be permanent fixtures.

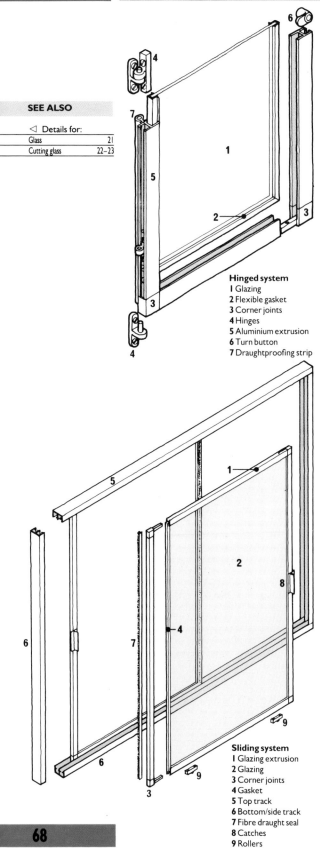

Hinged system
1 Glazing
2 Flexible gasket
3 Corner joints
4 Hinges
5 Aluminium extrusion
6 Turn button
7 Draughtproofing strip

Sliding system
1 Glazing extrusion
2 Glazing
3 Corner joints
4 Gasket
5 Top track
6 Bottom/side track
7 Fibre draught seal
8 Catches
9 Rollers

Types of glass used

Normally, 4mm ($\frac{5}{32}$in) glass is used on openable secondary glazing systems. For sliding windows no pane of glass should exceed 1.85 sq m (20 sq ft), and no more than 1.1 sq m (12 sq ft) should be used for side-hung, hinged sections.

When top-hung, each pane can be 1.65 sq m (18 sq ft). The height of each pane should not exceed 1.5m (5ft) or be greater than twice its width. Use toughened glass for low windows or those at risk from impact.

Hinged system

A hinged system incorporates an aluminium extrusion to form a frame for the glass or rigid plastic sheet. The glazing is seated in a flexible gasket. Corner joints fix the four sides together. Pivot hinges are fitted into the extrusion to make side-hung or top-hung units. A meeting rail section is also made for one frame to close against another where the window is too large to cover in one.

The hinged frames are fitted to the face of a wooden window frame and secured by turn buttons. A flexible draughtproofing strip is fixed to the back of the aluminium.

Sliding systems

A sliding system can operate horizontally or vertically. The horizontal version is normally used for casement windows and the vertical type for tall windows, such as double-hung sashes. You can get systems made of rigid plastic or aluminium.

Each pane of glass is framed by a lightweight extrusion which is jointed at the corners. The glass is sealed into its frame with a gasket. Two or more horizontally sliding panes can be used to suit the width of the window. They are held in a tracked frame, which is screwed to the window frame or the reveal. Fibre seals are fitted to the frame members to prevent draughts between the moving parts. The glazing is opened by a catch and each pane can be lifted out for cleaning.

A vertically sliding system uses a similar form of construction but incorporates catches to hold the panes open at any height.

Fixing a horizontally sliding system

Measure the width and height of your window opening. Buy the appropriate kit of parts to the nearest larger size. Cut the vertical and horizontal track members to size using a junior hacksaw (**1**) and fit the fibre seals in their grooves. Plug and screw them to the window reveal or the inside face of the window frame (**2**).

Following the manufacturer's instructions regarding tolerances, measure the opening for the glass and have it cut to size. Cut and fit the section of glazing frame, including the gasket. Join them together – usually with screw-fixed corner joints (**3**) but sometimes slot-together types – having inserted sliders and handles. Lift the glazing into the sliding tracks to complete the installation (**4**).

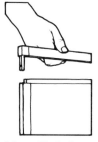

1 Cut track to length **2 Screw to the frame** **3 Assemble the frame** **4 Fit into the tracks**

Ventilation is essential for a fresh, comfortable atmosphere but it has a more important function with regard to the structure of our homes. It wasn't a problem when houses were heated with open fires, drawing fresh air through all the natural openings in the structure. With the introduction of central heating, insulation and draughtproofing, well-designed ventilation is vital. Without a constant change of air, centrally heated rooms quickly become stuffy and before long the moisture content of the air becomes so high that water is deposited as condensation (▷) – often with serious consequences. There are various ways to provide ventilation, some extremely simple, others much more sophisticated for total control.

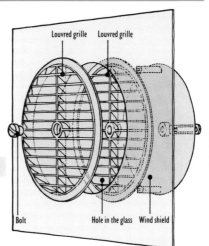

Initial considerations

Whenever you undertake an improvement which involves insulation in one form or another, take into account how it is likely to affect the existing ventilation. It may change conditions sufficiently to create a problem in those areas outside the habitable rooms so that damp and its side-effects develop unnoticed under floorboards or in the loft. If there is a chance that damp conditions might occur, provide additional ventilation.

Fitting a fixed window vent

Provide continuous 'trickle' ventilation by installing a simple fixed vent in a window. A well-designed vent allows a free flow of air without draughts, normally by incorporating a wind shield on the outside. It is totally reliable as there are no moving parts to break down or produce those irritating squeaks which are a feature of wind-driven fans.

Have a glazier cut the recommended size of hole in the glass (▷), then place one louvred grille on each side, clamping them together with the central fixing bolt. Bolt the clear plastic windshield to the outer grille.

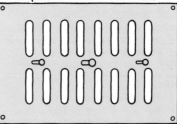

The components of a fixed window vent

Ventilating a fireplace

An open fire needs oxygen to burn brightly. If the supply is reduced by thorough draughtproofing or double glazing, the fire smoulders and the slightest downdraught blows smoke into the room. There may be other reasons why a fire burns poorly, such as a blocked chimney for example, but if it picks up within minutes of partially opening the door to the room, you can be sure that inadequate ventilation is the problem.

The most efficient and attractive solution is to cut holes in the floorboards on each side of the fire and cover them with a ventilator grille. Cheap plastic grilles work just as well, but you may prefer brass or aluminium for a living room. Choose a 'hit-and-miss' ventilator, which you can open and close to seal off unwelcome draughts when the fire is not in use. Cut a hole in a fully fitted carpet and screw the grille on top.

If the floor is solid, your only alternative is to fit a grille over the door to the room. An aperture at that height will not create a draught because cold air will disperse across the room and warm as it falls slowly.

Ventilating an unused fireplace

An unused fireplace that has been blocked by brickwork, blockwork or plasterboard should be ventilated to allow air to flow up the chimney to dry out penetrating damp or condensation. Some people believe a vent from a warm interior aggravates the problem by introducing moist air to condense on the cold surface of the brick flue. However, so long as the chimney is uncapped, the updraught should draw moisture-laden air to the outside. An airbrick cut into the flue from outside is a safer solution but it is more difficult to accomplish and quite impossible if you live in a terraced house. Furthermore, the airbrick would have to be blocked should you want to re-open the fireplace at a later date.

To ventilate from inside the room, leave out a single brick, form an aperture with blocks, or cut a hole in the plasterboard used to block off the fireplace. Screw a face-mounted ventilator over the hole or use one that is designed for plastering in. The thin flange for screw-fixing the ventilator to the wall is covered as you plaster up to the slightly protruding grille.

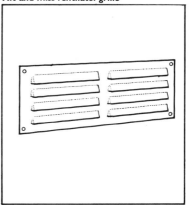

Face-mounted grille for ventilating a fireplace

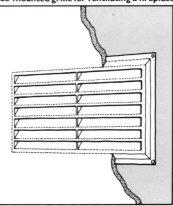

Hide the fixings of a grille with plaster

Hit-and-miss ventilator grille

VENTILATING BELOW FLOORS

Perforated openings known as airbricks are built into the external walls of a house to ventilate the space below suspended wooden floors. Without them, there's a strong possibility of dry rot developing in the timbers (◁), so check their condition regularly: they occasionally become clogged with leaves or earth piled against them, and rubble left by a builder may cover an airbrick on the inside. Clear a blocked airbrick as soon as you discover it, and if the original ones are inadequate replace them with new, larger ones.

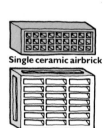
Single ceramic airbrick

Double plastic airbrick

Checking out the airbricks

Ideally there should be an airbrick every 2m (6ft) along an external wall but in many buildings there is less provision for ventilation with no ill-effects. Sufficient airflow is more important than the actual number of openings in the wall.

Floor joists that span a wide room are supported at intervals by low sleeper walls of brick. Sometimes they are perforated to facilitate an even airflow throughout the space. In other cases, the builder merely left gaps between sections of solid wall. This method of constructing sleeper walls can lead to pockets of still air in corners where draughts never reach. Even when all existing airbricks are open, dry rot can break out in those areas which never receive an adequate change of air.

If there appears to be likely dead areas under your floor, particularly if there are signs of damp or mould growth, fit an additional airbrick in a wall nearby.

Old ceramic airbricks do get broken and are often ignored because it has no detrimental effect on the ventilation. However, even a small hole provides access for vermin. Don't be tempted to block the opening, even temporarily, but replace the broken airbrick with a similar one of the same size. Choose from single- or double-size airbricks made in ceramic or plastic.

Installing or replacing an airbrick

Use a masonry drill to remove the mortar and a cold chisel to chop out the brick you are replacing. You may have to cut some bricks to install a double-size vent. Having cut through the wall, spread mortar on the base of the hole and along the top and both sides of the new airbrick. Push it into the opening, keeping it flush with the face of the brickwork. Repoint the mortar to match the profile used on the surrounding wall (◁).

Ventilating the space below a suspended floor
The illustration, right, is a cross-section through a typical cavity wall structure with a wooden floor suspended over a concrete base. A house with solid brick walls is ventilated in a similar way.
1 Airbrick
2 Sleeper wall built with staggered bricks to allow air to circulate
3 Joists and floorboards suffer from rot caused by poor ventilation

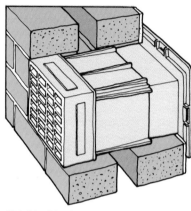

BRIDGING A CAVITY WALL

If you build an airbrick into a cavity wall, bridge the gap with a plastic, telescopic unit, which in turn is mortared into the hole from both sides. If necessary, a ventilator grille can be screwed to the inner end of the telescopic unit. Where an airbrick is inserted above the DPC (◁), you must fit a cavity tray over the telescopic unit to stop water running to the inner leaf of the cavity wall (◁).

Airbrick with telescopic sleeve
Bridge a cavity wall with this type of unit.

Cavity tray
A cavity tray sheds any moisture which penetrates the cavity above the unit to the outer leaf of the wall. It is necessary only when the airbrick is fitted above the DPC.

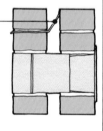

VENTILATING AN APPLIANCE

A flued, fuel-burning appliance must have an adequate supply of air to function efficiently and safely. If any alteration or improvement interferes with that supply you must provide alternative ventilation. If you plan to block a vent, change the window or even install an extractor fan in the same room as the appliance, consult a professional fitter. He will tell you whether the alteration is advisable, what type and size of vent to install, and where it should be positioned for best effect. An appliance with a balanced flue (◁) draws air directly from outside the house, so will be unaffected by internal alterations.

VENTILATING THE ROOF SPACE

When insulating the loft first became popular as an energy-saving measure, householders were recommended to tuck insulant right into the eaves to keep out draughts. What people failed to recognize was the fact that the free flow of air is necessary in the colder roof space to prevent moisture-laden air from the dwelling below condensing on the structure. Inadequate ventilation can lead to serious deterioration: wet rot develops in the roof timbers and water drops onto the insulant, eventually saturating the material and rendering it ineffective as insulation. If water builds up into pools, the ceiling below becomes stained and there is a risk of short-circuiting the electrical wiring in the loft. Efficient ventilation of the roof space, therefore, is an absolute necessity for every home.

Ventilating the eaves

Regulations for new housing insist on ventilation equivalent to continuous openings of 10mm (⅜in) along two opposite sides of a roof with a pitch (slope) of 15 degrees or more. If the pitch is less than 15 degrees, ventilation should amount to the equivalent of 25mm (1in) continuous openings. It makes sense to follow similar recommendations when refurbishing older houses.

The simplest method of ventilating a standard pitched roof is to fit round soffit vents made with integral insect screens. The spacing is determined by the size of opening provided by the particular vent. Push the vents into openings cut with a hole saw.

If the opening at the eaves is likely to be restricted by insulation, insert a plastic or cardboard eaves vent between each pair of joists. Push the vent into the angle between the rafters and the joists with the ribbed section uppermost. Vents can be cut to length with scissors for an exact fit. When installing blanket or loose-fill insulation (▷), push it up against the vent.

Slate and tile vents

Certain types of roof construction do not lend themselves to ventilation from the eaves only, but the structure can be ventilated successfully by replacing strategic tiles or slates with specially designed roof vents. A range of colours and shapes is available to blend with various roof coverings.

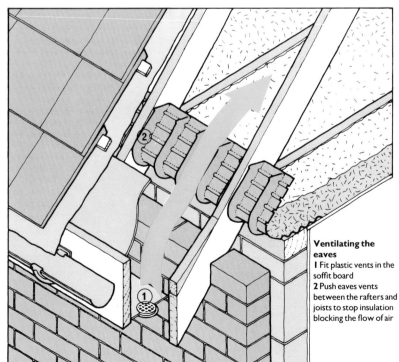

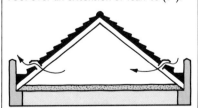

Ventilating the eaves
1 Fit plastic vents in the soffit board
2 Push eaves vents between the rafters and joists to stop insulation blocking the flow of air

WHEN ROOF VENTS ARE ESSENTIAL

Eaves-to-eaves ventilation is the standard method for keeping the roof space dry, but there are times when it is essential to fit tile or slate vents to draw air through the roof space.

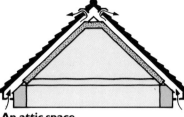

An attic space
If you insulate the slope of your roof you must provide a minimum 50mm (2in) airway between the insulant and the roof covering. It may be possible to fit eaves vents plus slate or tile vents near the ridge.

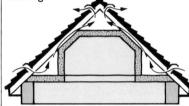

A room in the roof
Where a room is built into the attic, fit vents near the eaves and ridge to draw air through the narrow spaces over the sloping ceiling.

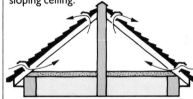

A fire or party wall
A solid wall built across the loft prevents eaves-to-eaves ventilation. Fit slate or tile vents to ventilate each side of the wall independently. Use the same arrangement to ventilate a mono pitch roof over an extension or lean-to (▷).

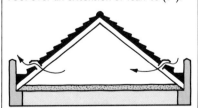

A roof with parapet
The roofs of older houses with parapets can be ventilated with vents positioned at a low level to give effective eaves-to-eaves ventilation.

Soffit vent

Slate/tile vents
Roof vents are made to resemble typical roof coverings.

Double Roman tile vent

Slate vent

Double pantile vent

FITTING ROOF VENTS

Installing a slate or tile vent

If you are contemplating having your roof replaced, ask the roofing contractor to incorporate vents at the same time, otherwise install vents in an existing roof yourself. Fitting instructions for individual models will vary in detail but the description below for a double pantile vent demonstrates the principle.

As each vent must be placed between rafters, select the approximate position of a vent and remove enough tiles (◁) to expose the felt and to locate the heads of the nails holding the tile-support battens to the roof. The nails indicate the position of the rafters.

Centre the template supplied by the manufacturer between the rafters and mark the position of the hole on the roof felt by scratching the corners with a knife (**1**). Cut the diagonals of the opening and bend back the flaps (**2**). Cut a slit in the felt, 100mm (4in) above and centred on the opening, and insert the tail of the undercloak of the vent (**3**), then align the edge of the undercloak with the opening (**4**). Plug the extension sleeve onto the underside of the cowl (**5**), then insert the sleeve into the hole in the felt and nail the cowl to the support batten above the opening (**6**). Replace the surrounding tiles, using hook clips to hold them in position (◁).

The undercloak
The undercloak fits beneath the vent cowl to prevent water running through the opening cut in the roof felt.

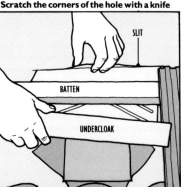

1 Scratch the corners of the hole with a knife

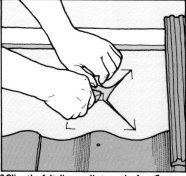

2 Slice the felt diagonally to make four flaps

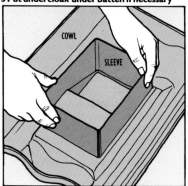

3 Put undercloak under batten if necessary

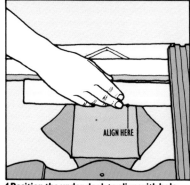

4 Position the undercloak to align with hole

5 Plug sleeve onto the bottom of the vent cowl

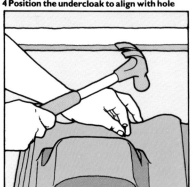

6 Nail the vent to the tile batten

CALCULATING AIR FLOW CAPACITY

All vents positioned near the eaves should provide the equivalent of 10 or 25mm (⅜ or 1in) continuous gap depending on the pitch of the roof. The ones near the ridge should provide air-flow to suit the construction of a particular roof.

Divide the specified airflow capacity of the vent you wish to use into the recommended continuous gap to calculate how many vents you will need. If in doubt, provide slightly more ventilation than is indicated.

Place eaves vents in the fourth or fifth course of slates or tiles. Position the higher vents a couple of courses below the ridge. Space all vents evenly along the roof to avoid any areas of 'dead' air.

CLEARING THE OPENING

When replacing certain tiles or slates with a vent it may be necessary to cut through a tile support batten to clear the opening. Nail a short length of batten above and below the opening to provide additional support.

Because roofing slates overlap each other by a considerable amount, you will have to cut away the top corners of the lower slates.

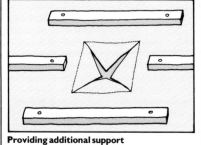

Providing additional support
Cut a tile batten that obstructs a hole, then place battens above and below to support the vent.

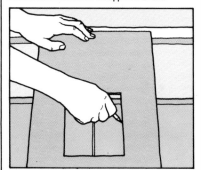

Marking slates that obstruct the hole
When slates cover the opening for a vent, use the template to mark the corners, then remove them.

FITTING AN EXTRACTOR FAN

Kitchens and bathrooms are particularly susceptible to problems of condensation so it is especially important to have a means of efficiently expelling moisture-laden air along with unpleasant odours. An electrically driven extractor fan will freshen a room faster than relying on natural ventilation and without creating uncomfortable draughts.

Positioning an extractor fan

The best place to site a fan is either in a window or on an outside wall but its exact position is more critical than that. Stale air extracted from the room must be replaced by fresh air, normally through the door leading to other areas of the house. If the fan is sited close to the source of replacement air it will promote local circulation but will have little effect on the rest of the room.

The ideal position would be directly opposite the source as high as practicable to extract the rising hot air (**1**). In a kitchen, try to locate the fan adjacent to the cooker but not directly over it. In that way, steam and cooking smells will not be drawn across the room before being expelled (**2**). If the room contains a flued, fuel-burning appliance, you must ensure there is an adequate supply of fresh air at all times or the extractor fan will draw fumes down the flue. The only exception is an appliance with a balanced flue, which takes its air directly from the outside.

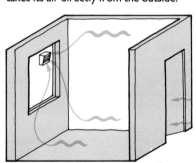

1 Fit extractor opposite replacement air source

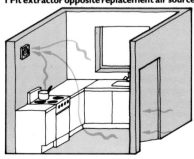

2 Place extractor near a cooker in a kitchen

Types of extractor fan

Many fans have integral switches but, if not, a switched connection unit can be wired into the circuit (▷) when you install the fan. Some models incorporate built-in controllers to regulate the speed of extraction and timers to switch off the fan automatically after a certain time. Fans can be installed in a window and some, with the addition of a duct, will extract air through a solid or cavity wall (see illustrations below). Choose a fan with external shutters that close when the fan is not in use, to prevent backdraughts.

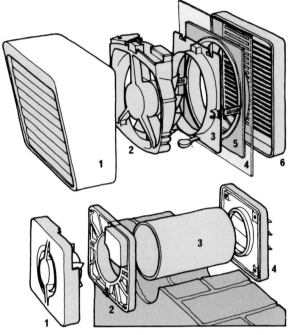

Window-mounted fan
1 Inner casing
2 Motor assembly
3 Interior clamping plate
4 Glass
5 Grille clamping plate
6 Exterior grille

Wall-mounted fan
1 Motor assembly
2 Interior backplate
3 Duct
4 Exterior grille

SEE ALSO	
Details for: ▷	
Switched connection unit	78
Condensation	10–11
Cooker hoods	75

Choosing the size of a fan

The size of a fan, or to be accurate, its capacity, is determined by the type of room in which it is installed and the volume of air it must move.

A fan installed in a kitchen must be capable of changing the air completely ten to fifteen times per hour. A bathroom requires fifteen to twenty air changes per hour and a WC, ten to fifteen. A living room normally requires four to six changes per hour, but fit a fan with a slightly larger capacity in a smoky environment.

To calculate the capacity of the fan you require, find the volume of the room (length x width x height) then multiply that figure by the recommended number of air changes per hour. Choose a fan which is capable of the same or slightly higher capacity.

CALCULATING THE CAPACITY OF FAN FOR A KITCHEN

SIZE			
Length	**Width**	**Height**	**Volume**
3.35m (11ft)	3.05m (10ft)	2.44m (8ft)	24.93m³ (880ft³)

AIR CHANGES		
Per hour	**Volume**	**Fan capacity**
15	24.93m³ (880ft³)	374m³ per hour (13,200ft³)

FITTING A WALL-MOUNTED UNIT

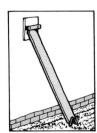

Metal detector
Detect buried pipes or cables by placing a hired electronic sensor against the plaster

Satisfy yourself that there is no plumbing or wiring buried in the wall by looking in the loft or under the floorboards (see left). Make sure there are no drain pipes or other obstructions.

Cutting the hole
Wall-mounted fans are supplied with a length of plastic ducting for inserting in a hole which you must cut through the wall to the outside. Plot the centre of the hole and draw its diameter on the inside of the wall. Use a long-reach masonry drill (◁) to bore a central hole right through. To prevent the drill breaking out brickwork or rendering on the outside, hold a stout plywood panel against the wall and wedge it with a scaffold board supported by stakes driven into the ground (**1**).

Before cutting the brick, drill holes close together around the inner edge of the hole. With a cold chisel, cut away the plaster using the holes as a guide, then continue to cut away the brickwork (try to avoid debris falling inside a cavity wall). When you reach the centre of the wall, remove the panel then use the same technique to finish the hole from the outside face.

1 Hold panel with plank

Fitting the fan
Most wall fans are fitted in a similar manner, but check specific instructions beforehand. Separate the components of the fan, then attach a self-adhesive foam sealing strip to the spigot on the backplate to receive the duct (**2**).

Insert the duct in the hole so that the backplate fits against the wall (**3**). Mark the length of the duct on the outside, allowing sufficient to fit the similar spigot on the outer grille. Cut the duct to length with a hacksaw. Reposition the backplate and duct to mark the fixing holes on the wall. Drill and plug the holes then feed the electrical supply cable into the backplate before screwing it to the wall. Stick a foam sealing strip inside the spigot on the grille. Position it on the duct then mark, drill and plug the wall fixing holes. Use a screwdriver to stuff scraps of loft insulation between the duct and the cut edge of the hole, then screw on the exterior grille (**4**).

If the grille does not fit flush with the wall, seal the gap with mastic. Wire the fan according to the manufacturer's recommendations. Attach the motor assembly to the backplate.

2 Seal plate spigot

3 Insert duct in hole

4 Screw-fix grille

FITTING AN EXTRACTOR FAN

Installing a fan in a window
An extractor fan can only be installed in a fixed window. If you wish to fit one in a sash window, it's necessary to secure the top sash in which the fan is installed and fit a sash stop (◁) on each side of the window to prevent the lower sash damaging the casing of the fan should it be raised too far.

To install an extractor fan in an hermetically sealed double glazing system, ask the manufacturer to supply a special unit with a hole cut and sealed around its edges to receive the fan. Some manufacturers supply a kit which adapts a fan for installing in a window with secondary double glazing. It allows the inner window to be opened without dismantling the fan.

Cutting the glass
Every window-mounted fan requires a round hole to be cut in the glass. The size is specified by the manufacturer. It is possible to cut a hole in an existing window but stresses in the glass will sometimes cause it to crack, and while the glass is removed for cutting there is always a security risk, especially if you decide to take it to a glazier. All things considered, it is advisable to fit a new pane: it's easier to cut and can be installed immediately the old one has been removed.

Cutting a hole in glass is not easy (◁) and it may be more economical in the long run to order it from a glazier. You'll need to supply exact dimensions, including the size and position of the hole. Use 3mm glass for a pane that does not exceed 0.2 sq m (2 sq ft) in area. Cut a larger pane from 4mm glass.

Installing the fan
The exact assembly may vary but the following sequence is a typical example of how a fan is installed in a window. Take out the existing window pane and clean up the frame, removing traces of old putty and retaining sprigs. Fit the new pane, with its hole pre-cut, as for fitting window glass (◁).

From outside, fit the exterior grille by locating its circular flange in the hole (**1**). Attach the plate on the inside, which clamps the grille to the glass. Tighten the fixing screws in rotation to achieve a good seal and an even clamping force on the glass (**2**). Screw the motor assembly to the clamping plate (**3**). Wire up the fan following the maker's instructions. Fit the inner casing over the motor assembly (**4**).

WARNING
Never make electrical connections until the power is switched off at the consumer unit (◁).

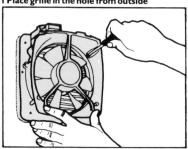

1 Place grille in the hole from outside

2 Clamp the inner and outer plate together

3 Screw the motor assembly to the plate

4 Attach the inner casing to cover the assembly

INSTALLING A COOKER HOOD

Window- and wall-mounted fans are designed for overall room extraction, but the ideal way to tackle steam and greasy cooking smells from a cooker is to mount a specially designed extracting hood directly over it.

Where to mount the cooker hood

Mount a cooker hood between 600mm (2ft) and 900mm (3ft) above the hob or about 400 to 600mm (1ft 4in to 2ft) above an eye-level grill. Unless the manufacturer provides specific dimensions, mount a hood as low as possible within the recommended tolerances.

Depending on the model, a cooker hood may be cantilevered from the wall or, alternatively, screwed between or beneath kitchen cupboards. Some kitchen manufacturers produce a cooker hood housing unit, which matches the style of the cupboards. Opening the unit operates the fan automatically. Most cooker hoods have two or three speed settings and an inbuilt light fitting to illuminate the hob or cooker below.

Installing trunking

When a cooker hood is mounted on an outside wall, air is extracted through the back of the unit into a straight duct passing through the masonry. If the cooker is situated against another wall, it is possible to connect the hood with the outside by means of fire-resistant plastic trunking. Straight and curved components plug into each other to form a continuous shaft running along the top of the wall cupboards.

Plug the female end of the first trunking component over the outlet spigot fitted to the top of the cooker hood. Cutting them to fit with a hacksaw or a tenon saw, run the rest of the trunking, making the same female-to-male connections along the shaft. Some trunking is printed with airflow arrows to make sure each component is orientated correctly: if you were to reverse a component somewhere along the shaft, air turbulence might be created around the joint, reducing the effectiveness of the extractor. At the outside wall, cut a hole through the masonry for a straight piece of ducting and fit an external grille (See opposite).

Fitting a cooker hood

Recycling and extracting hoods are hung from wall brackets supplied with the machines. Screw fixing points are provided for attaching them to wall cupboards. Cut a ducting hole through a wall as for a wall-mounted fan (See opposite). Wire a cooker hood following the maker's instructions.

RECIRCULATION OR EXTRACTION?

The only real difference between one cooker hood and another is the way it deals with the stale air it captures. Some hoods filter out the moisture and grease before returning the freshened air to the room. Other machines dump stale air outside through a duct in the wall, just like a conventional wall-mounted extractor fan.

Because the air is actually changed, extraction is the more efficient method but it is necessary to cut a hole through the wall and, of course, the heated air is lost – excellent in hot weather but rather a waste in winter. Cooker hoods which recycle the air are much simpler to install but never filter out all the grease and odours, even when new. It is essential to clean and change the filters regularly to keep the hood working at peak efficiency.

SEE ALSO

Details for: ▷
Condensation   10–11

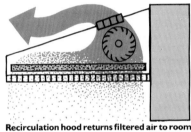

Recirculation hood returns filtered air to room

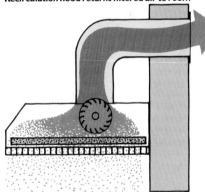

Extraction hoods suck air outside via trunking

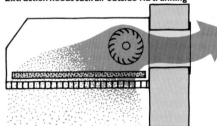

Alternatively, air is extracted through ducting

◁ **Running trunking outside**
When a cooker is placed against an inside wall, run plastic trunking from the extractor hood along the top of wall-hung cupboards.

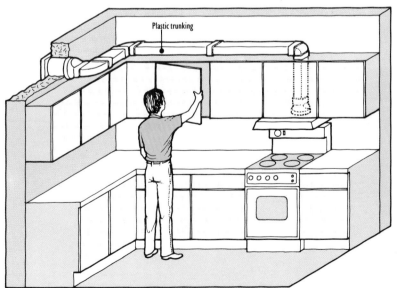

Plastic trunking

Most people naturally think of ventilating a room in hot weather when it is most likely to be unpleasantly stuffy. In all but the hottest of summers it can be achieved simply by opening windows or using a mechanical fan. However, ventilation is equally important in the winter – but we keep our weatherstripped windows closed against draughts, promoting an unhealthy atmosphere and a rise in condensation.

Using a conventional extractor fan solves the problem, but at the cost of throwing away heated air. A heat exchanging ventilator is the answer: two centrifugal fans operate simultaneously, one to suck in fresh air from the outside while the other extracts the same volume of foul air from inside. The system is perfectly balanced so the unit works efficiently even in a totally draughtproofed room.

The benefits of a heat exchanger

An important plus for heat exchanging ventilators is that heat loss is cut to a minimum by passing the extracted air through a series of small vents sandwiched between similar vents containing cold outside air moving in the other direction. Most of the heat is transferred to the fresh air so it is blown into the room warmed.

If you want to install a heat exchanger in a kitchen or bathroom, check that the unit is suitable for steamy atmospheres

and greasy air.

Heat exchangers are either wall-mounted much like a standard extractor fan, or flush-mounted so that the cabinet housing the unit is set into the wall. A slim, grilled casing only is visible from inside. The former version is relatively simple to fit but more obtrusive. To install a flush-mounted unit, it is necessary to cut a large rectangular hole through the wall and, in some cases, line it with timber.

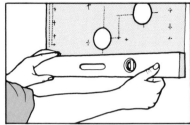

1 Position template to mark ducts and fixings

2 Bore holes for ducting with a hired core drill

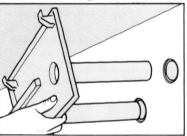

3 Pass both ducts through the wall

4 Plug gaps around ducts with insulation

5 Seal edge of duct covers with mastic

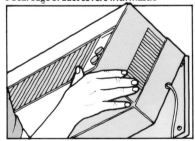

6 Fit the ventilator unit on the inside

How the ventilator functions
Hot, stale air is drawn into the unit (**1**) by a centrifugal fan (**2**). It is passed via the heat exchanger (**3**) to the outside (**4**). Fresh air is sucked from outside (**5**) by fan (**6**) to be warmed in the heat exchanger and passed into the room (**7**).

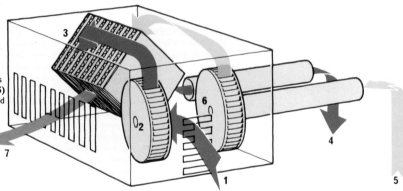

Fitting a wall-mounted ventilator

With the aid of a spirit level, use the manufacturer's template to mark the position of the ventilator on the wall, including the centre of both ducts (**1**). Locate the unit high on the wall but with at least 50mm (2in) clearance above and to the sides.

The ducting is likely to be narrower than that used for standard extractor fans, so it may be possible to use a hole saw or core drill to cut through the masonry (◁). Wedge a board outside to prevent the masonry breaking out (◁) then drill a pilot hole centred on each duct. Use a long-reach masonry drill to bore right through the wall. Angle the drill downwards five degrees so that the ducting will slope to drain condensation to the outside.

Locate the centre of the core drill in

the pilot hole then drill halfway through the wall before removing the board from the outside to continue there towards the inside (**2**). Use a hacksaw to cut the plastic ducting to a length equalling the depth of the wall plus 8mm (⅜in). Use aluminium flashing tape (◁) to hold ducting to both spigots on the rear of the wall-mounting panel. Insert the ducting into the drilled holes (**3**), push the panel against the wall and screw-fix it.

Outside, plug the gaps round the ducting with glass-fibre insulation (**4**) and screw the weather covers over the ends of the ducts. Seal around the edges of both covers with mastic (**5**). Fit the main unit to the mounting panel on the inside wall (**6**) and wire up to a fused connection unit nearby (◁).

VENTILATORS

Mounting a flush ventilator

Most flush-mounted ventilators require a wooden frame to line a hole cut through the wall. Make it to the dimensions supplied by the manufacturer of the ventilator. Construct the frame with lap joints or pinned and glued butt joints, and draw round it to mark the position of the hole on the wall. Apply a timber preservative.

Use a masonry drill to bore a hole through the wall at each corner then chop along the plaster between them with a bolster chisel. Drill further holes around the perimeter of the hole and chop out the masonry with a cold chisel. Remove whole bricks by drilling out the mortar joints. After cutting halfway through the wall, finish the hole from the outside.

The ventilator must be angled downwards a few degrees towards the outside to drain away condensation. If this angle is built into the ventilator unit, the wooden lining can be set flush with the wall, otherwise tilt it fractionally within the brickwork. Wedge the lining in place and fix it with masonry nails or fit wallplugs for fixing screws. Measure the diagonals to make sure the lining is square in the hole.

Fit the main unit in the liner and screw it to the front edge of the timber. Fit the front grille onto the unit. Outside, seal the junction between the unit and the lining with mastic. Do exactly the same where the timber frame meets the brickwork.

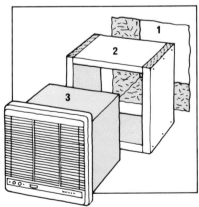

Mounting a flush ventilator
Cut a rectangular hole through the masonry (**1**). Construct a softwood liner using butt joints at the corners (**2**). Screw the ventilator (**3**) to the front edge of the liner.

FITTING A UNIT IN A TIMBER-FRAME WALL

Decide on the approximate position of the unit then locate the studs (▷) to align one with the side of the unit. Mark the rectangle for the wooden lining then drill through the plasterboard at the corners. Cut out the rectangle with a padsaw. Cut and remove the wall insulation. Working from outside, cut away the external cladding along the same lines. Saw off any studs obstructing the hole flush with its edges. Make and fit a timber lining as for a solid wall. Nail it to the studs and install the unit.

Mounting a ventilator liner ▷
Fit a timber liner in the wall close to the ceiling where it will be in the best position to extract hot rising air.

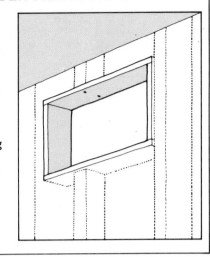

DEHUMIDIFIERS CONTROL CONDENSATION

To combat condensation you can remove the moisture-laden air by ventilation or warm it so that it can carry more water vapour before it becomes saturated. An alternative measure is to extract the water itself from the air using a dehumidifier. This is achieved by drawing air from the room into the unit and passing it over cold coils upon which water vapour condenses and drips into a reservoir. The cold, but now dry air is drawn by a fan over heated coils before being returned to the room as additional convected heat.

The process is based on the refrigeration principle that gas under pressure heats up and when the pressure drops, so does the temperature of the gas. In a dehumidifier, a compressor delivers pressurized gas to the 'hot' coils, in turn leading to the larger 'cold' coils, which allow the gas to expand. The cooled gas returns to the compressor for recycling.

A dehumidifier for domestic use is built into a cabinet which resembles a large hi-fi speaker. It contains a humidistat, which automatically switches on the unit when the moisture content of the air reaches a predetermined level. When the reservoir is full, the machine shuts down to prevent overflow and an indicator lights up to remind you to empty the water in the container.

When installed in a damp room, a dehumidifier will extract excess moisture from the furnishings and fabric of the building in a week or two. After that it will monitor the moisture content of the air to maintain a stabilized atmosphere. A portable version can be wheeled from room to room, where it is plugged into a standard wall socket.

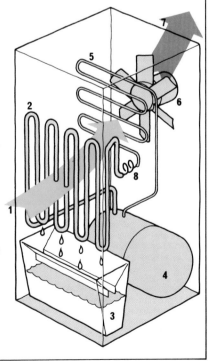

The working components of a dehumidifier ▷
The diagram illustrates the layout of a typical domestic dehumidifier.
1 Incoming damp air
2 Cold coils
3 Water reservoir
4 Compressor
5 Hot coils
6 Fan
7 Dry warm air
8 Capillary tube where gas expands

77

GLOSSARY OF TERMS

Aggregate
Particles of sand or stone mixed with cement and water to make concrete, or added to paint to make a textured finish.

Architrave
The moulding around a door or window.

Alkali-resistant primer
A primer used to prevent the alkali content of some building materials attacking subsequent coats of paint.

Arris
The sharp edge at the meeting of two surfaces.

Balanced flue
A ducting system which allows a heating appliance, such as a boiler, to draw fresh air from, and discharge gases to, the outside of a building.

Batt
A short cut length of glass- or mineral-fibre insulant.

Batten
A narrow strip of wood.

Blind
To cover with sand.

Blown
To have broken away, as when a layer of cement rendering has parted from a wall.

Buttercoat
The top layer of cement render.

Casing
The timber lining of a doorway.

Cavity tray
A sloping surface built in the void of a cavity wall to shed water to the outer leaf.

Cavity wall
A wall of two separate masonry skins with an airspace between them.

Chamfer
A narrow flat surface on the edge of a piece of wood – it is normally at an angle of 45 degrees to adjacent surfaces. *or* To plane the angled surface.

Chase
A groove cut in masonry or plaster to accept pipework or an electrical cable. *or* To cut such grooves.

Consumer unit
A box situated near the meter which contains the fuses or MCBs protecting all the circuits. It also houses the main isolating switch which cuts the power to the whole building.

Core drill
A hollow drill bit tipped with saw teeth to cut large-diameter holes in masonry.

Damp-proof course
A layer of impervious material which prevents moisture rising from the ground.

Damp-proof membrane
A layer of impervious material which prevents moisture rising through a concrete floor.

DPC
See Damp-proof course

DPM
See Damp-proof membrane

Drip groove
A groove cut or moulded in the underside of a door or windowsill to prevent rainwater running back to the wall.

Efflorescence
A white powdery deposit caused by soluble salts migrating to the surface of a wall or ceiling.

Elbow fitting
A plumbing joint designed to change direction in a pipe run.

Expansion pipe
An expansion pipe drains into the storage cistern in the roof to allow for expansion of heated water in the hot-water cylinder and to vent air from the plumbing system. Also known as vent pipe.

External wall insulation
Thermal insulating material which is fixed to the outer surface of a house in order to prevent transmission of heat to the outside.

Fascia board
Strip of wood which covers the ends of rafters and to which external guttering is fixed.

Flaunching
A mortared slope around a chimney pot or at the top of a fireback.

Footing
A narrow concrete foundation for a wall.

Frass
Powdered wood produced by the activity of woodworm.

Furring strips
Parallel strips of wood fixed to a wall or ceiling to provide a framework for attaching panels.

Fused connection unit
A device for permanently connecting electrical cable to flex – or sometimes another cable – from an appliance. The unit is protected by a cartridge fuse. Connection units sometimes incorporate a switch.

Hardcore
Broken bricks or stones used to form a sub-base below foundations, paving etc.

Jamb
The vertical side member of a door or window frame.

Key
To abrade or incise a surface to provide a better grip when gluing something to it.

Ladder stay
A device which holds the top of a ladder away from a wall so that one can service overhanging guttering, for instance, without having to lean back and possibly overbalance.

Lath and plaster
A method of finishing a timber-framed wall or ceiling. Narrow strips of wood are nailed to the studs or joists to provide a supporting framework for plaster.

Long-reach masonry drill
A drill designed to bore holes in brick, stone or concrete blocks but it incorporates an extended shank to allow the operative to reach into confined spaces or drill through thick walls.

Marking gauge
A woodworking tool designed to scribe a line parallel to a straight edge.

Mastic
A non-setting compound used to seal joints.

Microporous
Used to describe a finish which allows timber to dry out while protecting it from rainwater.

Mono-pitch roof
A roof which slopes in one direction only.

Mortar board
A 1m (3ft) square of exterior-grade plywood upon which mortar or plaster is mixed with water.

Mullion
A vertical dividing member of a window frame.

Oxidize
To form a layer of metal oxide as in rusting.

Paint stripper
A chemical which softens old paint so that it can be removed from a surface.

Paint system
The layers of paint applied to a surface, consisting of primer, undercoat and top coat.

Pallet
A wooden plug built into masonry to provide a fixing point for a door casing.

Panelling
Strips of solid timber or man-made boards used to line the face of a wall as a decorative finish.

Plugging chisel
An all-metal chisel with a flat narrow bit (tip) for cutting out old pointing between bricks or blocks.

PTFE
Polytetrafluorethylene – used to make tape for sealing threaded plumbing fittings.

Render
A thin layer of cement-based mortar applied to exterior walls to provide a protective finish. Sometimes fine stone aggregate is embedded in the mortar. *or* To apply the mortar.

Sash stop
A metal stud screwed into the framework of a sliding sash window to prevent it being opened further than required for ventilation.

Scaffold tower
A tall structure built from metal poles and scaffold boards to provide access to the upper sections of a house for decoration and repair.

Scratchcoat
The bottom layer of cement render.

Screed
A thin layer of mortar applied to give a smooth surface to concrete etc. *or* A shortened version of screed batten.

Screed batten
A thin strip of wood fixed to a surface to act as a guide to the thickness of an application of plaster or render.

Stile
A vertical side member of a door or window sash.

Switched connection unit
See Fused connection unit.

Work platform
A structure constructed from scaffolding or trestles and boards used to gain access to walls and ceilings for decoration or repair. See also Scaffold tower.

INDEX

Page numbers in *italics* refer to photographs and illustrations